Thin Films and Nanostructures

Waveguide Spectroscopy of Thin Films

Volume 33

Serial Editors

VLADIMIR AGRANOVICH
Institute of Spectroscopy
Russian Academy of Sciences
Moscow, Russia

DEBORAH J. TAYLOR
Freescale Semiconductor
Austin, Texas

Honorary Editors

MAURICE H. FRANCOMBE
Department of Physics
and Astronomy
Georgia State University
Atlanta, Georgia

STEPHEN M. ROSSNAGEL
IBM Corporation,
T. J. Watson Research Center
Yorktown Heights, New York

ABRAHAM ULMAN
Alstadt-Lord-Mark Professor
Department of Chemistry
Polymer Research Institute
Polytechnic University
Brooklyn, New York

Editorial Board

DAVID L. ALLARA
Pennsylvania State University

ALLEN J. BARD
University of Texas, Austin

FRANCO BASSANI
Scuola Normale Superiore, Pisa

MASAMICHI FUJIHIRA
Tokyo Institute of Technology

GEORGE GAINS
Rensselaer Polytechnic Institute

PHILLIP HODGE
University of Manchester

JACOB N. ISRAELACHIVILI
University of California
Santa Barbara

MICHAEL L. KLEIN
University of Pennsylvania

HANS KUHN
MPI Gottingen

JEROME B. LANDO
Case Western Reserve University

HELMUT MOHWALD
University of Mainz

NICOLAI PLATE
Russian Academy of Sciences

HELMUT RINGSDORF
University of Mainz

GIACINTO SCOLES
Princeton University

JEROME D. SWALEN
International Business
Machines Corporation

MATTHEW V. TIRRELL
University of Minnesota,
Minneapolis

CLAUDE WEISBUCH
Ecole Politechnique, Paris

GEORGE M. WHITESIDES
Harvard University

ANVAR ZAKHIDOV
University of Texas at Dallas

Recent volumes in this serial appear at the end of this volume

Thin Films and Nanostructures

Waveguide Spectroscopy of Thin Films

Alexander V. Khomchenko

*National Academy of Science of Belarus
Institute of Applied Optics
Mogilev 212793, Belarus*

VOLUME 33

2005

ELSEVIER
ACADEMIC
PRESS

Amsterdam – Boston – Heidelberg – London – New York – Oxford – Paris
San Diego – San Francisco – Singapore – Sydney – Tokyo

ELSEVIER B.V.	ELSEVIER Inc.	ELSEVIER Ltd	ELSEVIER Ltd
Radarweg 29	525 B Street, Suite 1900	The Boulevard, Langford	84 Theobalds Road
P.O. Box 211	San Diego	Lane, Kidlington	London
1000 AE Amsterdam	CA 92101-4495	Oxford OX5 1GB	WC1X 8RR
The Netherlands	USA	UK	UK

© 2005 Elsevier Inc. All rights reserved.

This work is protected under copyright by Elsevier Inc., and the following terms and conditions apply to its use:

Photocopying

Single photocopies of single chapters may be made for personal use as allowed by national copyright laws. Permission of the Publisher and payment of a fee is required for all other photocopying, including multiple or systematic copying, copying for advertising or promotional purposes, resale, and all forms of document delivery. Special rates are available for educational institutions that wish to make photocopies for non-profit educational classroom use.

Permissions may be sought directly from Elsevier's Rights Department in Oxford, UK: phone (+44) 1865 843830, fax (+44) 1865 853333, e-mail: permissions@elsevier.com. Requests may also be completed on-line via the Elsevier homepage (http://www.elsevier.com/locate/permissions).

In the USA, users may clear permissions and make payments through the Copyright Clearance Center, Inc., 222 Rosewood Drive, Danvers, MA 01923, USA; phone: (+1) (978) 7508400, fax: (+1) (978) 7504744, and in the UK through the Copyright Licensing Agency Rapid Clearance Service (CLARCS), 90 Tottenham Court Road, London W1P 0LP, UK; phone: (+44) 20 7631 5555; fax: (+44) 20 7631 5500. Other countries may have a local reprographic rights agency for payments.

Derivative Works

Tables of contents may be reproduced for internal circulation, but permission of the Publisher is required for external resale or distribution of such material. Permission of the Publisher is required for all other derivative works, including compilations and translations.

Electronic Storage or Usage

Permission of the Publisher is required to store or use electronically any material contained in this work, including any chapter or part of a chapter.

Except as outlined above, no part of this work may be reproduced, stored in a retrieval system or transmitted in any form or by any means, electronic, mechanical, photocopying, recording or otherwise, without prior written permission of the Publisher.

Address permissions requests to: Elsevier's Rights Department, at the fax and e-mail addresses noted above.

Notice

No responsibility is assumed by the Publisher for any injury and/or damage to persons or property as a matter of products liability, negligence or otherwise, or from any use or operation of any methods, products, instructions or ideas contained in the material herein. Because of rapid advances in the medical sciences, in particular, independent verification of diagnoses and drug dosages should be made.

First edition 2005

ISBN: 0-12-088515-8

∞ The paper used in this publication meets the requirements of ANSI/NISO Z39.48-1992 (Permanence of Paper). Printed in the United States of America.

To the closest person,

without whose attention,

help and patience,

this work would have been impossible,

to my wife Valentina

Contents

Waveguide Spectroscopy of Thin Films. xi
Foreword. xiii
Acknowledgments. xv

Introduction 1

Chapter 1. Interaction of Light with Matter
1.1. Power States in Solids. 5
1.2. Macroscopic Aspects of Solids. 12

Chapter 2. Spectroscopy of Optical Guided Modes
2.1. Waveguide Properties of Thin Films and Surface Layers . . . 21
2.2. Dispersion Curves and the "Cut-Off" Condition 25
2.3. Input of Radiation into Waveguide by the Prism Coupler . . 27
2.4. Excitation of Guided Light Modes and Measuring their
       Parameters. 29
2.5. Optical Losses in Waveguides 31
2.6. Leaky and Plasmon Modes in Thin-Film Structures 37
2.7. Fabrication of Waveguiding Structures. 39

Chapter 3. New Applications of the m-Line Technique for Studying Thin-Film Structures
3.1. Recording the Spatial Distribution of the Light-Beam
       Intensity 41
3.2. Techniques and Setup for the Measurement of Light
       Beam Intensity and its Spatial Distribution. 61
3.3. Spatial Distribution of the Light Beam Intensity Reflected
       from the Prism Coupler and Measurements Thin-Film
       Parameters. 70

Chapter 4. Spatial Fourier Spectroscopy of Guided Modes: Measuring Thin-Film Parameters

4.1. Fourier Transform Applications for Studying the Spatial Distribution of the Reflected Light Beam Intensity 77
4.2. Studying the Properties of Waveguiding Films 83

Chapter 5. Characterization of Thin Films by Prism Coupling of Leaky Modes

5.1. Basic Concepts and Instrumentation 98
5.2. Determinations of Waveguiding Film Parameters 103
5.3. Leaky Modes in Thin-Film Structures 106
5.4. Determinations of Parameters of Metal Films and Surface Layers of Bulk Metal by the Plasmon Mode Excitation Technique 110

Chapter 6. Measurements of Absorption Spectra of Thin Films

6.1. Absorption Optical Spectrophotometry: Possibilities and Limitations 114
6.2. Instrumentation of Waveguide Spectroscopy of Thin Films....................................... 115
6.3. Special Features of Absorption Spectra Recording by Waveguide Spectroscopy and their Processing.......... 119
6.4. Measurement of Absorption Spectra by the Fourier Spectroscopy of Guided Modes..................... 120

Chapter 7. Applications of Waveguide Spectroscopy Techniques in Sensor Systems

7.1. Integrated-Optics Sensors and their Features........... 128
7.2. Gas Thin-Film Sensors 134
7.3. Physical Origin of Processes on Surfaces of Thin-Film Sensors 139
7.4. Evaluation of Adlayer Parameters 146
7.5. Sensors with Recording of the Light Beam Reflection Coefficient...................................... 147
7.6. Waveguide Microscopy of Thin Films 151

Chapter 8. Optical Nonlinearity in Thin Films at Low-Intensity Light
8.1. Optical Nonlinearity and Thin-Film Nonlinear Constant
   Measurement Techniques 156
8.2. New Method of Determination of Nonlinear Constants
   of Films...................................... 164
8.3. Nonlinear Optical Properties of Azo-Dye Doped
   Polymeric Films................................ 169
8.4. Optical Nonlinearity in Semiconductor Films 172
8.5. Nonlinear Absorption in Semiconductor-Doped Glasses ... 185

Chapter 9. Optical Nonlinearity in Multilayer Structures
9.1. Features of Optical Properties of Multilayer Structures 191
9.2. Interface Effects on the Nonlinear Optical Properties of
   Thin Films.................................... 197
9.3. New Potentialities for Studying the Thin-Film
   Structures 202

References 207
Subject Index 219

A.V. Khomchenko
Waveguide Spectroscopy of Thin Films

In this book, new methods of studying the linear and nonlinear optical properties of thin films are presented. These techniques are based on the principles of the spatial Fourier spectroscopy of light beam reflected from a prism-coupling device of tunnel excitation of guided light modes in thin-film structures. Measurement techniques for determining absorption coefficient, refractive index and thickness of the dielectric, semiconductor or metallic films are considered. Examples of applications of the described methods for determining the adlayer thickness and impurity concentrations in the surroundings and also for the reconstruction of the refractive index profile of gradient planar structures are presented. The technique of measurement of nonlinear optical constants of thin films and the results of studying optical nonlinearity of semiconductor and dielectric films including the low-dimensional structures are stated.

This book is therefore, for the specialists in the fields of integrated and thin-film optics, for PhD students and students of other related disciplines.

Foreword

The intense development in the new devices and in the applications of thin films in optics and electronics have stimulated investigations to improve the existing methods of measuring thin-film parameters and to develop new measurement techniques. Surface optics and nonlinear optics of thin-film structures are essential in studying the physical properties of thin films, because problems of spectroscopy of these thin films with low optical losses arise while studying the electronic state distribution in the band gap of thin-film materials. The general tendency of the miniaturization of micro- and opto-electronic elements, and the properties of the devices are now determined by the surface and interface qualities. In essence, the evolution of nonlinear optics of thin films, caused by the wide spectrum of the applications of thin films, has stimulated the need for further studies of the nonlinear medium properties and for the development of new nonlinear optics methods to investigate surfaces and interfaces.

Usage of waveguide propagation phenomena of light in thin firms or surface layers makes it possible to create non-destructive diagnostic techniques with a high resolution that maintains the integrity and the quality of the investigated sample. It also allow one to directly measure the film's refractive index. The existing methods of direct measurements of the thin-film and surface layer parameters require either the destruction of the investigated sample or results in less accurate characterization of the thin films. In addition, the existing integrated-optics methods do not allow one to determine the absorption coefficient of thin films. The problem of measurement of small optical losses has always been quite difficult, especially when one deals with the determination of the losses of dielectric thin films in the visible range of the spectrum. Although there have been numerous attempts to solve this task, they have not led to any significant results. But with waveguide spectroscopy techniques of thin film, these problems have been resolved.

The term "*waveguide spectroscopy*" implies a way of obtaining information about thin-film properties. It includes the excitation of guided modes in thin-film structures, measurement of the mode characteristics and determination of film spectral–optical parameters by using measured data. These measurement techniques are based on the principles of spatial Fourier spectroscopy of the reflected light beam when the guided mode is excited in thin-film structures by a prism coupler, i.e., they are based on the recording of the spatial spectrum of the intensity of the light beam reflected from a prism coupler. In this approach, the advantages of the resonant and interference measurement techniques are incorporated, and consequently, the methods considered here have high sensitivity and resolution.

In this book, the basic concepts of waveguide spectroscopy of thin films, special features of measurement techniques, specific devices, and some aspects of studying thin films and media properties are described. The book is the result of many years of study of thin-film properties by using integrated-optics methods by the author. Some of the results were arrived at in cooperation with Dr. A.B. Sotsky and Dr. A.A. Romanenko, who provided the theoretical background for the possibility of the determination of the complex permittivity of thin films by the prism-coupling techniques.

Acknowledgments

This book is the result of my work during several years in the field of thin-film physics. I would like to mention the names of my academic teachers Professors P. Apanasevich, A. Goncharenko and V. Red'ko.

It would be necessary to express my sincere gratitude to the colleagues of the Institute of Applied Optics of National Academy of Sciences, co-authors of joint works, results of which are used in this book, and for their fruitful cooperation and also the employees of the laboratory of optic-electronic devices for their long-term support in this field. I am grateful to Prof. A.A. Afanas'ev, Dr. A.I. Voytenkov and Dr. A.B. Sotsky for their remarks on the separate chapters of the book, which made better the contents of some of the questions raised. Special gratitude is expressed to Professors A.M. Goncharenko and S.V. Gaponenko, who read the book in the manuscript stage and made useful remarks on its contents.

I am also grateful also to E. Glazunov, Yu. Lebedinsky, A. Pravotorov, I. Primak and A. Romanenko for providing the software for measurements and data processing, and special thanks are expressed to E. Khomchenko and D. Khomchenko for their help in preparation of the book.

This work was supported by Belarus Republic Fund of Fundamental Researches.

Introduction

The progress in modern optics of thin films is caused as a result of investigations and applications of the optical phenomena and effects, which take place in such structures. The application of new effective measurement methods allows one to find such effects, and the results of these studies expand our representations about the origin and mechanisms of the interaction of light with matter. The study of solid surfaces and thin-film properties in turn promotes not only discoveries of fundamental character, but also initiates the development of new opto- and microelectronic technologies. Therefore, it is important to improve the existing methods of investigations of film and surface layer property and also develop new ones.

A great number of thin-film structures, used in optics and microelectronics, which differ by their optical, electrical and geometric parameters, stimulate the development of various measurement techniques, which take into account the specificity and peculiarities of the investigated samples. A well-known measurement method of thin-film parameters is the interferometry method, which can measure the refractive index and the absorption coefficient with an accuracy of 10^{-6} [1, 2] and 10^{-4} (with a value of absorption coefficient of about 2×10^{-3}) [3], respectively. Traditional refractometer methods are the simplest and they are applied for the measurement of the refractive index on the surface, whereas for studying bulk samples, the goniometric methods are used [1, 4]. Ellipsometry has been traditionally used for measuring optical parameters and film thickness of non-absorbant layers [7], a thorough review of which can be found in Ref. [8]. But these methods are not adequate for measuring the parameters of absorptive films and of metals, in particular [9]. Direct measurement of the refractive index and the absorption coefficient is not possible in a number of cases. For this reason indirect methods of measuring the dielectric constant are used in spectroscopy. In particular, the measurement of the reflected light spectrum (by varying the incidence angle of the light beam, at a different polarization or in surroundings with different refractive indexes, etc.) is quite often performed, instead of the measuring absorption coefficient [10]. But these methods have an important restriction: The results remain unaffected by the changes in the reflection factor caused by great changes in optical parameters of the film. Since the measurement accuracy of the reflection factor is restricted, errors of the reflectivity measurements can cause substantial errors in the calculation of optical parameters.

A number of problems may arise in the case of determination of small optical losses, especially when considering small loss measurements in thin films. Traditional absorption spectroscopy [11] does not provide information about the characteristics of low absorptive samples. The application of such special methods as multireflection measurements [12, 13] increases the sensitivity of measurements, but it does not completely solve the problem. The application of laser radiation with high spectral density for those of diagnostics of thin films greatly extend the possibilities of low loss measurement. The development of thermooptic [14] and photoacoustic methods [15, 16] allows one to record small losses, but the sensitivity of these methods are limited by the magnitude of $\alpha d \approx 10^{-4}$ [17], where α and d are the absorption coefficient and the films thickness, respectively. By laser spectroscopy based on absorption of laser radiation as a result of multireflection in the high-quality resonator, the absorption coefficients are measured to about 10^{-4}–10^{-9} cm^{-1} [18]. But such techniques are only applicable in recording the narrow absorption lineshapes, which are typical for gases. But for solids the sensitivity of the absorption coefficient measurement decreases [19]. The latest measurement techniques also do not allow one to determine the absolute value of the absorption coefficient but rather compare it with a previous value [3]. Thus, the nondestructive methods of testing thin films are most promising [20]. These are the methods based on a resonant excitation of guided light modes by a prism coupler, and have been described in this book. The phenomenon of light propagation as a guided mode is useful for the investigation of the properties of thin films and surface layers. These features appear, for example, with the possibility to measure the refractive index of thin films with a thickness of 0.1 micrometer to tens of micrometers [21, 23]. At the same time, as has been shown in Ref. [28], the waveguide techniques are more accurate in comparison to other measurement techniques.

The results described in this book are the generalizations and extensions of the waveguide methods of investigation of the spectral-optical properties of thin films. The questions of interaction of light with matter, basic concepts of the macroscopic description of the interaction of electromagnetic radiation with solids are considered in Chapter 1.

The waveguide measurement techniques are based on the determination of the propagation constants of the guided modes excited in the structure under investigation. The mode propagation constant can be determined by recording the resonance minimum in the angular dependence of the light beam reflection factor when the guided modes are excited by a prism coupler [24–28]. But the only information about the angular location of the reflection minimum (position of dark m-line) is taken into account and it stipulates the determination of only the refractive index and the film thickness [29–32]. The optical losses (intrinsic attenuation of light in waveguides) are usually recorded in the additional experimental measurements of the mode attenuation with the propagation of light along the waveguide [33, 34]. These questions are considered in Chapter 2, where the terminologies are entered and the basic aspects of the propagation of the electromagnetic waves in thin-film structures are considered.

Chapter 3 is devoted to the study of the spatial distribution of laser beam intensity reflected from a prism device of tunnel excitation of guided modes and its correlation with the optical and the geometrical parameters of thin films. The known methods of the measurement of thin-film parameters by the prism-coupling technique are based on the recording of the angular position of the dark m-lines only. At the same time the m-line has a "fine" structure, which contains information about the film properties.

In Chapter 4, a new method for studying the properties of thin-film structures is described. The method considers the recording of the spatial distribution of the reflected light beam intensity, well known as the waveguide methods. But the distinctive feature of this method is the manner in which the whole massif of the experimental data in the intensity distribution is used, including the contrast, width of all angle spectrum and its separate elements. All this as a whole allows the determination of the absorption coefficient of the film together with its thickness and refractive index.

Modes excited in thin-film structures can be guided [23–27], leaky [35, 36] or plasmon modes [37, 38]. In Chapter 5, we study optical waveguides, systems with thin-film covers in microelectronics (structures as SiO_2–SiO_xN_{1-x}–Si, SiO_2–Si), metal films or surface layers of bulk metals.

One of the most interesting applications of the waveguide spectroscopy methods, discussed in Chapter 6, is the measurement of the absorption spectrum of thin films in the spectral range of its transparency. We should mention the fact that the high sensitivity of the measurement technique allows one to determine the spectral absorption coefficient to $1\,cm^{-1}$ for films with thickness of about 0.1 µm.

Besides, the use of the peculiarities of the propagation of light modes in films and also the high sensitivity of these methods allow one to determine the monomolecular adlayer parameters and restore the refractive index distribution of two-dimension gradient samples. It also provides with the opportunity to create new types of sensors. The examples of the application of the structures based on the prism coupler as different types of waveguide sensors are considered in Chapter 7.

Under the influence of coherent radiation the parameters of a nonlinear medium undergo such insignificant changes that it can only be determined by highly sensitive methods. Methods of measuring the nonlinear refractive index and nonlinear absorption coefficient are considered in Chapter 8.

The study of the optical nonlinearity discovered in the semiconductors, dielectric films and multilayer structures are discussed in Chapter 9. These effects are recorded at light intensity $<0.1\,\text{W/cm}^2$.

Chapter 1
Interaction of Light with Matter

1.1. Power States in Solids 5
1.2. Macroscopic Aspects of Solids 12

By studying complex dielectric permittivity one can obtain some information about processes in solids taking place under the influence of light. Direct measurements of the absorption coefficient are possible over quite a wide spectral range. The spectral range, which is where the refractive index can be measured, is limited by the transmission band, which lies in an infrared spectral range for the majority of semiconductors, but for most oxide or soda-halogen materials, this band is in the infrared to ultraviolet range. On the other hand, in the transmission band of optical materials, one can record the absorption bands, where the absorption coefficient reaches values of 10^1–10^2 cm^{-1} in maxima at different concentrations of defects or impurities. But the measurement of such low absorption in thin films by traditional techniques is practically impossible.

Integrated-optics methods are actively applied for studying the properties of thin films and surface layers. This chapter therefore covers the basic notions characterizing the electromagnetic field in vacuum or in a medium, and discusses the general principles of interaction and propagation of electromagnetic fields in solids including thin-film structures. A discussion of the terminology and the minimum knowledge necessary for investigation of thin-film properties by waveguide methods can also be found in this chapter. This is also justified by the fact that there are a number of books dedicated to the investigation of optical properties of solids [40–46] and to spectroscopy techniques [47–52].

1.1. Power States in Solids

The measurement of the absorption spectrum is the most direct and perhaps also the simplest method of investigating the band structure of solids. By measuring light absorption in a given spectral range, one can obtain information about the distribution function of energy levels. As optical phenomena are related to transitions between different energy states and as these states affect the optical properties of materials, it would be appropriate to discuss at first as the electronic states, so their origin.

Optical properties of dielectrics, semiconductors and metals will be discussed in all the chapters of this book. Such a division of real materials is made on the basis of the value of electric conductivity, σ. At this conditional classification, important characteristics such as dependence of resistance or conductivity on temperature, structure and substance properties should also be taken into account. The difference between semiconductors and dielectrics has some conditional origin as they only differ with respect to the value of conductivity and activation energy. In order to create conductivity in semiconductors one needs to transform them to an excited state, while for metals, the conductivity state is their natural, unexcited state. Such an approach allows one to distinguish semiconductors from metals: an extrinsic influence weakly affects the metal conductivity. Peculiarities of material properties are defined by the electron state system of atoms forming solids. Conductivity in solids can be explained on the basis of energy representations. As it follows from quantum mechanics, in an isolated atom, the electrons occupy certain power levels (Figure 1.1(a)). With the formation of condensed materials, i.e. with reapprochement of atoms up to distances that are typical for substances in such conditions, the electrons will be in the strong field of the neighboring atoms. The interaction of electrons with other electrons and with surrounding atoms leads to essential changes in the energy level system of the electrons in an atom. The power levels of such electrons are split into an energy band or energy zone. The energy related to levels of such electron become smaller in comparison with energy levels of electron in isolated atom. The influence of a crystal lattice on electron motion in solids leads to the existence of the energy levels or even entire bands, which can be occupied by electrons. These allowed energy bands in the

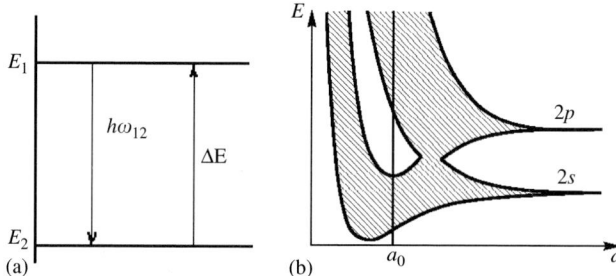

Fig. 1.1. Energy levels of isolated atom (a), their splitting and formation of energy bands at the change of distance between atoms [11], where a_0 is a typical distance for solids (b).

distribution of electrons over the quantum states are separated from each other by the forbidden energy ranges (the band gap).

The shape of the band defines electron properties. In real crystals, the motion of the electrons is affected by other electrons and neighboring atoms, which is why it is quite difficult to describe it. So we will consider only the qualitative description of this phenomenon.

The width of the energy band depends on the interatomic distance, and it increases exponentially with the decrease of this distance. We can draw the dependence of allowed energy bands on the lattice constant a of an atomic system (Figure 1.1(b)). Let us assume that the atomic system has one occupied and one empty level. At the reapprochement of atoms every level is split and its position is changed at the same time. If the crystal contains 10^{20} atoms, there is such an enormous quantity of individual quantum states that they form a continuous band of allowed energy levels. After the overlapping of upper and lower levels it is impossible to say whether the lower or upper band is formed from the lower or upper atom level, respectively. Every band is created as the superposition of atomic functions of both states. This phenomenon is typical for upper valence bands and for lower conduction bands of many crystals. Whether the solid will be a conductor or isolator depends on the structure of energy bands. The energy bands for semiconductor materials and for metals are schematically shown in Figure 1.2. There are two types of such bands: the lowest occupied band is usually referred to as the valence band and the highest unoccupied band is called the conduction band. The interval between the top of the valence band, E_v, and the bottom of the conduction band, E_c, is referred to as the band gap, $E_g = E_c - E_v$, and E_F is the chemical potential commonly referred to as the Fermi energy or the Fermi level. It is known that almost all electrons are situated in the valence band at the thermal equilibrium, i.e. they are concentrated and kept in certain places of a solid lattice.

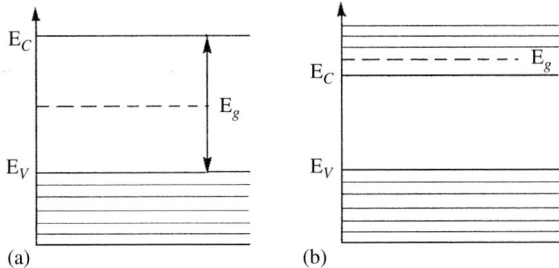

Fig. 1.2. Energy bands of (a) semiconductors and (b) metals.

A distribution of electron population $N(E)$ is given by the Boltzmann law

$$N_i = N \exp(-E_i/k_b T),$$

where N is the total number of molecules in the medium, N_i is the density of population at the level with energy E_i and k_b is the Boltzmann constant [53].

If electrons are imposed with additional energy from outside, then some electrons situated at the lower energy level transit to the higher energy level and $N_2 > N_1$. The imposition of additional energy on electrons leads to the generation of free electrons, which can move inside the semiconductor. This happens because some electrons concentrated in the valence band pass into the conductivity band. The positive centers or holes appear on the free places in the valence band. The holes and free electrons are the charge carriers in semiconductors. It is obvious, that in a gas, the electrons that occupy the upper energy level, lose energy and come back to the lower level because of collisions with other electrons. Free electrons in semiconductors collide with the atoms of lattice and other electrons, pass into the valence band and an "electron–hole" pair disappears. Sometimes, the transition to the lower energy level or valence band takes place without collisions. In these cases the energy lost by the electron is released as a photon. Such a process of irradiation is referred to as spontaneous radiation. A spectral frequency of such radiation is defined by the difference of energy levels equal to $(E_2 - E_1)$ or $(E_c - E_v)$, i.e. by the band gap value

$$\omega = (E_2 - E_1)/\hbar = E_g/\hbar,$$

where $\hbar$ is the Planck constant.

In the case of a local perturbance of the crystal field caused by impurity atoms, dot defects or dislocation the allowed states, related to the perturbation area, appear in the band gap, and the electrons are localized in the area of local perturbations. The power level relating to the localized state appears as a result of the detachment of the highest (lowest) level in the allowed band and its following transition into the band gap. The rest of the energy levels remain almost unchanged. If the concentration of local perturbation N increases, the average distance between them changes to $\sim N^{-1/3}$. The wavefunctions of localized states overlap with the increase in N and the levels are transformed into the band with the energy being changed quasi-continuously. For example, the localized states in the band gap appear at the implantation of impurity atoms into the material. A doping of crystals with a donor impurity leads to the appearance of a discrete level under the bottom of the conduction band, and the acceptor impurity, in turn, leads to the appearance of the level above the top of the valence band. The scheme of the impurity levels is depicted in Figure 1.3. A transition of electrons from the impurity donor level into the allowed band is equivalent to the

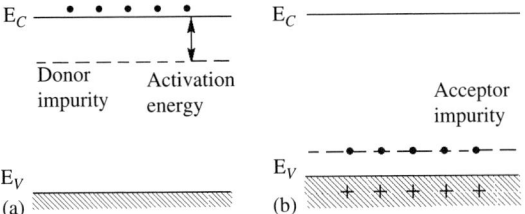

Fig. 1.3. Scheme of power levels for donor (a) and acceptor (b) impurities.

ionization of an impurity atom. Hence, the impurity activation energy should be equal to the ionization energy, which varies in the range of 4–24 eV for different materials. But during the implantation of an atom, as an impurity, its valence electrons interact with the atoms of the main substance. This leads to the reduction of the bond between the electron and the impurity atom.

Let us try to evaluate the position of the impurity level in the band gap. We can show that the impurity ionization energy is equal to

$$E_i = \frac{1}{2} \frac{Ze^4 m^*}{\hbar^2 \varepsilon^2}, \qquad (1.1.1)$$

where $\hbar$ is Planck's constant, ε, e, Z and m^* are the permittivity, electron charge, atomic number, and effective mass of electron in the media doped by the impurity, respectively.

The ionization energy can be expressed in electron-volts (eV); by substituting the numerical values of constants into Eq. (1.1.1) [53], one has

$$E_i = \frac{13.52 Z^2}{\varepsilon^2} (m^*/m) \quad (\text{eV}). \qquad (1.1.2)$$

Since $m^* < m$, where m is the electron mass and $\varepsilon^2 \approx 100$, then E_i values are in the range 0.05–0.1 eV.

The real materials contain a number of defects and impurities, which form the energy level system of the electron states. Despite the fact that the concentration of such impurities can be low, they play quite a significant role in the spectroscopy of solids, especially in the study of effects related to the absorption and irradiation of photons with energy lower than the optical band gap. One can find a more detailed description of defects in solids in Refs. [54–56]. The number of different types of defects is extremely large. Their classification often depends on the observable properties of defects we are interested in. If we take the origin of defects as a starting point, we can sort them into intrinsic and impurity defects [43]. The notion of impurity defects is obvious. There are defects caused by atoms, which are not contained in the main substance. Vacancies, atoms in intermodes can be intrinsic defects [57]. Geometrical properties of defects allow one to divide them into point defects related to some atoms, linear (e.g. dislocations), plane (e.g. glide plane) and volume (impurities and pores) [56]. In fact, the interface can also be considered as the plane defect, because the crystal periodicity is broken at the interface and the material structure is often changed near interfaces. Defects can capture or return electrons. In semiconductor theory such defects are called acceptors or donors. The name "trap" – for electron or hole – can be used for both types of defects depending on what the main capture process is. Defects can also be classified depending on their interaction with light. For example, if the energy level of the electron state relates to the defect in crystal lying in the band gap, then such defects are called color centers, because they lead to coloration of crystals, which are transparent in the visible spectral range. If the defect intensively captures excited electrons, leading to radiative recombination in the form of a luminescence, then such defects are called luminescence centers [56]. This classification can be extended. Attention should be focused on the conditional character of the defect classification, as the same defects can exhibit different behavior under different conditions.

The deep defect states appear in the absorption spectra in the form of sharp peaks. The position of these peaks is determined by the energy of the related energy levels of defects and peak intensity by the concentration of defects taking into account a degree of degeneracy of energy levels. Only the centers interacting with charge carriers due to the coulomb field can create quasi-continuous system of levels. Lower levels close to the continuum can be described as the Rydberg states:

$$E = -\text{Ry}/n_l^2, \quad (1.1.3)$$

where n_l is the main quantum number (energy is calculated from the bottom gap), and Ry is the Rydberg constant.

Every energy level with same value of n_l is degenerated $2n_l$ times over orbital and magnetic quantum numbers. The difference in energies between two neighboring states is equal to

$$dE/dn_l = 2Ry/n_l^3.$$

So the state density for highly excited states of centers with concentration N in volume V is equal [10] to

$$g(E) = Nn^2 \frac{dn}{dE} = NV \frac{n_l^3}{2Ry} = NV \frac{Ry^{3/2}}{2|E|^{3/2}} = \frac{m^{*3/2} e^6 NV^2}{2^{5/2} \varepsilon^3 \hbar^3 |E|^{5/2}}. \quad (1.1.4)$$

The $g(E)dE$ value is equal to the number of states allowed for electron with some energy and spin per volume unit in the energy range between E and

($E+dE$). Thus, the state density in crystals with impurities has a tail in the band gap (Figure 1.4). The similar scheme is depicted also in Ref. [60]. In addition to lower energy levels there are local levels situated at quite significant distance from the bottom of an energy band. As we will see, such levels are of great importance for describing non-equilibrium processes.

Owing to the finiteness of the crystal size and the presence of broken bonds in surface layers and for films – close to their interfaces, there are the surface states localized near interfaces. The respective energy levels are called the power levels of surface states. Their position inside the band gap is determined by the surface quality and surrounding media. The energy levels of surface states for some materials are given in Ref. [58]. The number of such states in crystal solids attains values of 10^{16} cm^{-2}, and, in general, the surface states have great influence on physical processes in solids. In reality, we come across polycrystalline and amorphous thin-film materials more often than single crystals. The application of periodicity notion for describing crystal materials is just a theoretical approach that provides a quantitative analysis or crystal properties [42]. Periodicity allows one to consider only the elementary lattice and one has to further use translation symmetry to obtain optoelectronic properties of materials. From the conceptual point of view, the short-range ordering is important for determining these properties of materials. Experiments on the diffraction of X-rays and electrons reveal the approximate identity of their nearest neighbors in amorphous and crystal silicon. The photoemission of electrons shows that the state density in crystalline and amorphous materials of the same composition are approximately of the same value [42].

As it is quite difficult to describe amorphous materials mathematically, in practice, the model series, which somehow explains experimental data, can

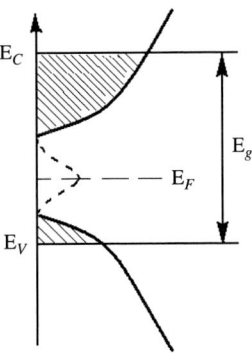

Fig. 1.4. The scheme of band structure of the glassy As_2S_3, shaded areas are the localized states.

be used. Every model should introduce a notion of the state density $g(E)$ as well as of the crystals. The power levels of these states can be either free or occupied. The shape of the function $g(E)$ can be determined from the experimentally measured data. This function is similar to the function of the electron distribution over energies for crystal materials, but a great number of localized states appear inside the band gap of non-crystalline materials [59]. The assumption that states near the valence band and conduction band in amorphous semiconductors are localized, i.e. they are capture centers, is confirmed experimentally. So, one can introduce quantities E_c^* and E_v^*, equal to the energy levels separating the energy bands.

These quantities are analogs of the bottom of the conduction band and of the top of the valence band in crystals, respectively. In this book the energy gap between these two levels will be represented as the band gap E_g by analogy with crystals. We should define one more feature of the conductive non-crystalline materials as well as of all defect structures. It appears as the fixing of the Fermi level E_F in the middle of the band gap (or near the E_v level) [58, 60] due to enormous amount of structural and surface defects.

So, the concept introduced before for crystalline solids can be applied for polycrystalline and amorphous materials as well. Some researchers draw a conclusion that lattice coordination is more important than long-range ordering [42], hence the concept of the state density $g(E)$ can be applied for the description of non-crystalline solids. If we know the dependence $g(E)$ it would be possible to determine optical characteristics such as the refractive index or the absorption coefficient, for example.

1.2. Macroscopic Aspects of Solids

Let us formulate the basic concepts characterizing the investigated optical medium from the viewpoint of phenomenological description of interaction of electromagnetic radiation with condensed materials. Such approach is correct not only for describing the influence of external electromagnetic fields on medium but also for the characterization of solids themselves. The International System of Units (SI) has been used throughout the book. The complex quantities are represented by bold face fonts and normal characters and the scalar quantities by italic font.

While analyzing effects, related to the propagation of light in thin-film structures we will often refer to the wave equation for isotropic and homogeneous medium. It can be easily obtained from the Maxwell equation [61]

$$\partial^2 \vec{E}/\partial x^2 - (1/c^2)\, \partial^2 \vec{E}/\partial t^2 = (1/c^2 \varepsilon_0)\, \partial \vec{j}/\partial t, \qquad (1.2.1)$$

where E is the electric field strength, ε_0 the permittivity of vacuum, c the speed of light in vacuum, and j the total current density of charges with the carrier concentration N.

It is supposed that the volume density of charge ρ does not depend on coordinates, and the medium is non-magnetic. The density of charge is the sum of the free charges (ρ_e) with concentration N_e and the bound charges (ρ_b) with concentration N_b. Hence, $\vec{j} = \vec{j}_f + \vec{j}_b$, and includes two components: $\vec{j}_f$ and $\vec{j}_b$ are the current density for free and bound charges, respectively. The conduction current $\vec{j}_f$ can be expressed as

$$\vec{j}_f = eN_e \bar{v} = eN_e(d\bar{x}/dt) = \partial \vec{P}/\partial t, \qquad (1.2.2)$$

where x is the shift of electron with charge e under the influence of electric field $\vec{E}$, $\bar{v} = d\bar{x}/dt$ the rate of free electron, and $P_f = eN_e \bar{x}$ the polarization of free charges.

Here $\vec{P}$ is the dipole moment of the substance volume unity $\vec{P} = N\vec{p}$, where $\vec{p} = e\vec{r}$ is the dipole moment of one particle, and $\vec{r}$ is the distance between positive and negative charges.

Analogous expressions can be written for bound electrons and $\vec{P}_b = e\vec{\xi}N_b$ is the polarization of bound charges (electrons), where $\vec{\xi}$ is the shift.

It is obvious that such notions as polarization, shift and other quantities associated with them are used only when free charges are affected by the gradient field $E(t)$, because $\bar{x}$ and P_f have definite and finite values only in this case.

Hence

$$\vec{P} = P_f + P_b. \qquad (1.2.3)$$

We consider here only linear materials. The shifts $\vec{\xi}$ and $\bar{x}$ in such media are described by linear differential equations. Hence polarization $\vec{P}$ depends linearly on the electric field [62]

$$\vec{P} = \varepsilon_0 \chi \vec{E} \qquad (1.2.4)$$

Here χ is the dielectric susceptibility, which consists of two components,

$$\chi = \chi_b + \chi_f, \qquad (1.2.5)$$

where $\chi_b = N_b \alpha_b$, $\chi_f = N_e \alpha_f$, and α_b and α_f are polarizabilities of the medium for bound and free electrons, respectively.

Since $\vec{P} = \vec{P}_f + \vec{P}_b$ and $\vec{j} = \partial \vec{P}/\partial t$ then $\partial \vec{j}/\partial t = \partial^2 \vec{P}/\partial t^2$. The right part of the wave equation (1.2.1) can be written as the second derivative of the total polarization of the substance:

$$\partial^2 \vec{E}/\partial \bar{x}^2 - (1/c^2)\partial^2 \vec{E}/\partial t^2 = (1/c^2 \varepsilon_0)\partial^2 \vec{P}/\partial t^2. \qquad (1.2.6)$$

Owing to the fact that the dipole moment is equal to the product of the electron charge and the proper shift, one can say that polarization P is the volume concentration of dipole moments. So, the polarization is the surface density of charges (free and bound) that appear on the surface due to the electric field. Using these relationships one can derive the following expression from the wave equation:

$$\partial^2 \vec{E}/\partial \bar{x}^2 - (n^2/c^2) \times \partial^2 \vec{E}/\partial t^2 = 0, \qquad (1.2.7)$$

where n is the refractive index of the medium.

We should take into account the fact that the electric field $\vec{E}$ and the vector of electric induction $\vec{D}$ in isotropic medium are related by the expression

$$\vec{D} = \varepsilon \varepsilon_0 \vec{E} = \varepsilon_0 \vec{E} + \vec{P}, \qquad (1.2.8)$$

then $\varepsilon = 1 + \vec{P}/\varepsilon_0 \vec{E}$, with ε is the relative permittivity of medium.

From (1.2.8), the refractive index is determined by the expression

$$\varepsilon = n^2 = 1 + \vec{P}/\varepsilon_0 \vec{E} \qquad (1.2.9)$$

or

$$\varepsilon = n^2 = 1 + \chi = 1 + \chi_b + \chi_f, \qquad (1.2.10)$$

and from (1.2.4) follow expressions for susceptibilities:

$$\chi_b = \vec{P}_b/\varepsilon_0 \vec{E} = e\vec{\xi} N_b/\varepsilon_0 \vec{E},$$

$$\chi_f = \vec{P}_f/\varepsilon_0 \vec{E} = e\vec{x} N_e/\varepsilon_0 \vec{E}. \qquad (1.2.11)$$

The propagation of electromagnetic wave $\vec{E}(z,t)$ and its behavior in space are determined by the function $n(\omega)$, i.e. by dispersion or by dependence of the medium properties on the frequency of the incident radiation. Let us consider for simplicity a one-dimensional case and search for a solution of the wave equation (1.2.8) as a plane wave, propagating along the direction Z:

$$E = E_0 e^{i\omega t - k_0 z},$$

where $k_0 (= n\omega/c)$ and n are the wavenumber and the refractive index, respectively.

During wave propagation in a medium, a decrease in wave amplitude also takes place (the light absorption). One can get a simple and clear idea about the relation between dispersion and absorption of light from a classic model of interaction between the radiation and the medium. This model describes electrons as harmonic oscillators, performing forced oscillation in a field of the light wave $\vec{E}$. This model allows one to relate the refractive index of medium with its microscopic properties.

In order to determine the refractive index n and its dependence on frequency, the polarization P should be calculated. To determine P one needs to obtain the shifts ξ and x from the appropriate differential equations, which describe the motion of bound and free electrons affected by the harmonic field. For the bound electron this is the equation of harmonic oscillator with the fundamental frequency ω_0, mass m and damping factor δ:

$$d^2\xi/dt^2 + 2\delta d\xi/dt + \omega_0^2 \xi = (eE_0/m)e^{i\omega t}. \quad (1.2.12)$$

The motion equation for free electron that takes into account electron collisions with mean lifetime τ is given by

$$d^2x/dt^2 + (1/\tau)dx/dt = (eE_0/m)e^{i\omega t}. \quad (1.2.13)$$

Let us estimate the susceptibility χ_b, which is specified by the contribution of free electrons. The solution of Eq. (1.2.13) for free electron (for simplicity let $\delta = 0$) is

$$x = \frac{eE_0 e^{i\omega t}}{m(-\omega^2 + i\omega/\tau)} = \frac{-eE_0 e^{i\omega t}}{\omega^2 m(1 - i/\omega\tau)}.$$

The expressions for polarizability α_f and for susceptibility $\chi_f = N_e \alpha_f$ of free electrons can be written as

$$\alpha_f = -\frac{e^2/m\varepsilon_0}{\omega^2(1 - i/\omega\tau)},$$

$$\chi_f = -\frac{e^2 N_e}{m\varepsilon_0 \omega^2 (1 - i/\omega\tau)} = \frac{\omega_p^2}{\omega^2(1 - i/\omega\tau)}, \quad (1.2.14)$$

where $\omega_p^2 = e^2 N_e / m\varepsilon_0$ is the squared plasma frequency.

Making some simple transformations one can rewrite this equation as

$$\lambda_c^2 N_e r_e = \pi,$$

where $\lambda_c = 2\pi c/\omega_p$ is the critical wavelength.

Expression (1.2.14) is also written as

$$\chi_f = -\frac{\lambda_c^2}{\lambda_c^2(1 - i/\omega\tau)} = -\frac{\lambda^2(1 + i/\omega\tau)}{\lambda_c^2[1 + (\omega\tau)^{-2}]} \quad (1.2.15)$$

This equation leads to the conclusion that existence of free electrons causes the presence of the imaginary part of permittivity, and hence the refractive index also has non-zero imaginary part. If in Eq. (1.2.10) the value of χ_b can be neglected in comparison to unity, then

$$\varepsilon = \mathbf{n}^2 \cong 1 - \frac{\lambda^2(1 + i/\omega\tau)}{\lambda_c^2[1 + (\omega\tau)^{-2}]}. \quad (1.2.16)$$

Sometimes, this relationship is easier to write in the following way:

$$\varepsilon = \mathbf{n}^2 \cong 1 - \frac{e^2 N_e}{m\varepsilon_0 \omega^2 (1 - i/\omega\tau)}.$$

In this case the following expression for the refractive index is written as

$$\mathbf{n} \cong \pm \left[1 - \frac{e^2 N_e}{m\varepsilon_0 \omega^2 (1 - i/\omega\tau)}\right]^{1/2}. \quad (1.2.17)$$

So, peculiarities of the refractive index define such quantities as critical wavelength λ_c, frequency ω or wavelength $\lambda = 2\pi c/\omega$ and the mean time between collisions τ, i.e. these magnitudes define whether the refractive index will be a complex, pure imaginary or real quantity.

In order to clarify the physical sense of the complex refractive index let us write out its real and imaginary parts:

$$\mathbf{n} = n + ik, \quad (1.2.18)$$

where k is the absorption coefficient (extinction coefficient).

The permittivity $\varepsilon(\omega) = \varepsilon'(\omega) + i\varepsilon''(\omega)$ is also a complex function of the radiation frequency. This is a fundamental property associated with the causality principle [63]. The introduction of two pairs of functions (ε' and ε''; n and k), are uniquely related to each other by the expressions

$$\varepsilon' = n^2 - k^2, \qquad \varepsilon'' = 2nk \quad (1.2.19)$$

This is caused by the fact that values of n and k are measured in experiments in most cases and the quantity $\varepsilon = \varepsilon' + i\varepsilon''$ is usually calculated theoretically by using microscopic parameters. By substituting n in the form (1.2.18) into the wave equation (1.2.7) and further solving shows that local amplitude $E_0(z)$ is exponentially decreased along the direction of the axis Z:

$$E = E_0 e^{-\beta z},$$

where $\beta = k\omega/c = 2\pi k/\lambda$ is the attenuation coefficient of the electric field.

The energy transferred by electromagnetic wave is determined by the expression

$$\vec{S} = [\vec{E} \times \vec{H}] = c^2 \varepsilon_0 [\vec{E} \times \vec{B}], \quad (1.2.20)$$

where $\vec{S}$ is the Poynting vector that is functionally related to the volume density of electric energy w_v^E, magnetic energy w_v^B and to the Joule losses P_v^{J-L} in a medium (in the case when the medium is a conductor) by the expression

$$\text{Div } \vec{S} = \frac{\partial}{\partial t}(w_v^E + w_v^H) - P_v^{J-L}.$$

Here $w_v^E = \varepsilon_0 E^2/2$, $w_v^B = \mu_0 H^2/2 = (1/\varepsilon_0 c^2)(B^2/2)$, $P_v^{J-L} = \bar{j}E$ [61], H is the intensity of the magnetic field, B the magnetic induction, μ_0 the magnetic conductivity of vacuum, and μ_0 and ε_0 are constants related by the expression $1/\varepsilon_0\mu_0 = c^2$.

The Poynting vector module characterizes the energy flux density of the electromagnetic waves. The time averaging of S is the density of the electromagnetic radiation or the light intensity I. The light intensity at the distance z from the surface is described by the well-known Buger–Lambert's law:

$$I = \varepsilon_0\{E^2\}_t = (1/2)\varepsilon_0 E^2 \exp(-2\beta z) = I_0 \exp(-\alpha z), \qquad (1.2.21)$$

where $\{E^2\}_t$ is the time-averaged square of the field.

For the harmonic field with $E = E_0 \cos \omega t$ we have $\{E^2\}_t = E_0^2/2$ and $I_0 = I(0) = \varepsilon_0 E_0^2/2$ is the initial value of the light intensity. The quantity $\alpha = 2\beta$ is the coefficient of intensity attenuation or the absorption coefficient related to the imaginary part k of the complex refractive index n by the expression

$$\alpha = 4\pi k/\lambda. \qquad (1.2.22)$$

Here k is the quantity that will be used further. It will be often called the absorption coefficient, but we should remember that the relationship between the quantity k and the absorption coefficient α is as given in (1.2.22).

Let us estimate the dielectric susceptibility χ_b defined by the contribution of electrons bounded with atoms. The solution of Eq. (1.2.16) gives the following result:

$$\xi = (eE_0/m)e^{i\omega t}/(\omega_0^2 - \omega^2 + i\delta\omega/m).$$

The following relationships can be written for the polarizability α_b and susceptibility χ_b:

$$\alpha_b = (e^2/\varepsilon_0 m)/(-\omega_0^2 + i\delta\omega/m),$$

$$\chi_b = N_b\alpha_b = (e^2 N_b/\varepsilon_0 m)/(-\omega^2 + \omega_0^2 + i\delta\omega/m).$$

The frequency ω_0 and the corresponding wavelength $\lambda_0 = 2\pi c/\omega_0$ are defined by the total energy of the main state of bound electrons. This energy is equal to the ionization energy of atoms. Setting it equal to 10 eV, for example, we will obtain $\lambda_0 = 0.124\,\mu\text{m}$. Moreover, the expression $\omega_b^2 = e^2 N_b/\varepsilon_0 m$ is similar to the well-known formula for plasma frequency. So, the expression for the refractive index of optical materials will be as follows:

$$\mathbf{n} = 1 + \frac{N_b e^2}{\varepsilon_0 m(\omega_0^2 - \omega^2 + i\delta\omega)}. \qquad (1.2.23)$$

The substitution of Eq. (1.2.18) into Eq. (1.2.23) allows one to obtain dependencies of n and k on frequency:

$$n = 1 + \frac{N_b e^2 (\omega_0^2 - \omega^2)}{2\varepsilon_0 m[(\omega_0^2 - \omega^2)^2 + \delta^2 \omega^2]}, \qquad (1.2.24a)$$

$$k = \frac{N_b e^2 \delta \omega}{2\varepsilon_0 m[(\omega_0^2 - \omega^2)^2 + \delta^2 \omega^2]}. \qquad (1.2.24b)$$

If the light frequency is close to the frequency of optical transition, then expressions (1.2.24a) can be rewritten as

$$n = 1 + \frac{N_b e^2 (\omega_0^2 - \omega^2)}{4\varepsilon_0 m[(\omega_0^2 - \omega^2)^2 + (\delta/2)^2]},$$

$$k = \frac{N_b e^2 \delta}{8\varepsilon_0 m[(\omega_0^2 - \omega^2)^2 + (\delta/2)^2]}. \qquad (1.2.25)$$

The dependencies of n and k on frequency, which is close to the fundamental frequency, are shown in Figure 1.5.

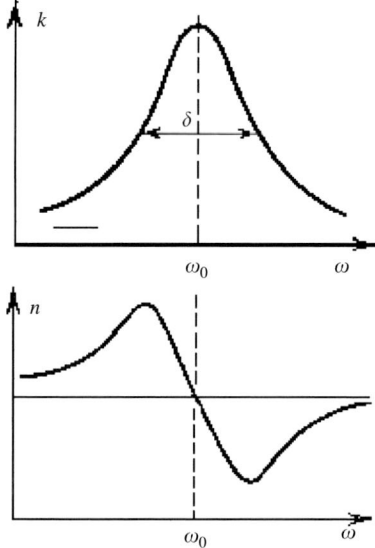

Fig. 1.5. Dependencies of n and k on frequency near to the fundamental frequency ω_0 of optical transitions.

The relationship between ε' and ε'' is described by the Kramers–Kronig dispersion relationships:

$$\varepsilon'(\omega) - 1 = \frac{2}{\pi} \text{vp} \int_0^\infty \frac{\omega' \varepsilon''(\omega)}{\omega'^2 - \omega^2} d\omega', \qquad (1.2.26a)$$

$$\varepsilon''(\omega) - \frac{4\pi\sigma}{\omega} = \frac{2\omega}{\pi} \text{vp} \int_0^\infty \frac{\varepsilon'(\omega') - 1}{\omega'^2 - \omega^2} d\omega'. \qquad (1.2.26b)$$

If we know only $\varepsilon''(\omega)$ in the full spectral range, the values of ε' can be determined with the help of expressions (1.2.26a). Owing to the singularity in the point $\omega = \omega'$ the integrals in (1.2.26b) are regarded as their principal values. The dispersion relation relating to the refractive index and the absorption coefficient is written as

$$n(\omega) - 1 \leqslant \frac{2}{\pi} \text{vp} \int_0^\infty \frac{\omega' k(\omega')}{\omega'^2 - \omega^2} d\omega. \qquad (1.2.27)$$

In the assumption of

$$\frac{a}{\pi} \text{va} \int_0^\infty \frac{\omega k(\omega)}{(\omega')^2 - \omega^2} d\omega = 0.$$

Expression (1.2.27) can be written as

$$n(\omega) - 1 = \frac{2}{\pi} \int_0^\infty \frac{\omega' k(\omega')}{(\omega')^2 - \omega^2} d\omega, \qquad (1.2.28)$$

where the singularity area is extracted. Note that at $\omega = \omega'$,

$$\lim \frac{\omega' k(\omega') - \omega k(\omega)}{(\omega')^2 - \omega^2} = \frac{k(\omega)}{2\omega} + \frac{1}{2} \frac{dk}{d\omega}.$$

In the calculation of the refractive index the region of integration is split into four ranges:

- from ∞ to ω_1, where $k(\omega)$ is described by the Drude model;
- from ω_1 to ω_p, where the harmonic oscillator model is used;
- in the frequency range from ω_p to ω_h the approach based on the polynomial approximation of the values $k(\omega)$, which are measured experimentally, is used; and
- in the range from $\omega = \omega_h$ to $\omega = \infty$, the asymptotic interpolation of $\varepsilon(\omega)$ is used for determining the k.

So, to determine the refractive index with the help of expression (1.2.28) one needs to calculate the following four integrals:

$$n(\omega) - 1 = I_{dm} + I_{sgm} + I_{exp} + I_{ae} \qquad (1.2.29)$$

While determining I_{dm}, the film parameters are calculated on the basis of Drude model. In this case ε can be considered as $\varepsilon = \varepsilon_1 + \varepsilon_2$, where ε_1 is the magnitude depending on the influence of free electrons, and ε_2 is the magnitude depending on the influence of bound electrons. The expression for $\varepsilon_1(\omega)$ is given by the well-known model of free electrons

$$\varepsilon_1(\omega) = 1 - \frac{\Omega_p^2}{\omega(\omega + iT_0)}, \quad (1.2.30)$$

and the function ε_2 is described by the Lorenz relationship

$$\varepsilon_2(\omega) = \frac{k}{z_j} \frac{f_i \omega_p^2}{(\omega^2 - \omega_j^2) + i\omega T_j}, \quad (1.2.31)$$

where $\Omega_p = (Ne^2/m\varepsilon_0)^{1/2}$.

In the calculation of I_{exp} the interpolation of the experimental dependence $k(\omega)$ is typically used in the form of

$$k = k_e + C_{1e}(\omega - \omega_e) + C_{2e}(\omega - \omega_e)^2 + C_{3e}(\omega - \omega_e)^3 + \cdots, \quad (1.2.32)$$

where ω is the current frequency, and the value k is measured at frequency ω_e. The coefficients C_e can be found from the following relationship:

$$C_{i,e} = \frac{dk}{d\omega}\bigg|_{\omega \to \omega_e}.$$

In the high-frequency range the approximation of I_{ae} is based on the analysis of the permittivity behavior:

$$\varepsilon'(\omega) = 1 - \frac{4\pi Ne^2}{m\omega^2}, \quad (1.2.33)$$

$$\varepsilon''(\omega) = \frac{A}{\omega^3},$$

where N is the total number of electrons, and $A = 1.42003 \times 10^9$ if ω is defined in eV.

Note that the negative value of susceptibility χ_f in its absolute value can be sufficiently greater than unity. This implies that the imaginary refractive index $n = ik$. Thus, in the case of the light propagation in such medium the sufficient attenuation of electromagnetic field occurs at a length comparable with wavelength λ and even at smaller lengths [64]. The values $k > 1$ in visible spectral range is not uncommon for metals [65].

Chapter 2
Spectroscopy of Optical Guided Modes

2.1. Waveguide Properties of Thin Films and Surface Layers. 21
    2.1.1. Reflection of the Plane Wave from Interface of Two Media 22
    2.1.2. Modes in Planar Waveguiding Structures 23
2.2. Dispersion Curves and the "Cut-Off" Condition 25
2.3. Input of Radiation into Waveguide by the Prism Coupler. 27
2.4. Excitation of Guided Light Modes and Measurement of Their Parameters. 29
2.5. Optical Losses in Waveguides. 31
    2.5.1. Dispersion Equations for the Imaginary Part of the Mode Propagation 32
    Constant. . . . .
    2.5.2. Measurement of Losses in Waveguides 34
2.6. Leaky and Plasmon Modes in Thin-Film Structures. 37
2.7. Fabrication of Waveguiding Structures 39

We will consider light propagation in thin-film structures in this chapter. The integrated-optics techniques of investigation of thin film and layer properties are based on the phenomenon of waveguide propagation of light inside them. A planar optical waveguide, being the basic element of integrated optics, is a dielectric layer on a suitable dielectric substrate. The thin-film (uniform) and gradient optical waveguide depend on the refractive index distribution across the guiding layer there are.

2.1. Waveguide Properties of Thin Films and Surface Layers

In a thin-film waveguide (Figure 2.1) the refractive index n_f of the guiding layer is greater than refractive index n_s of the substrate and the refractive index n_c of the surrounding medium. This condition is necessary to provide the waveguiding regime, i.e. for the existence of optical guided modes in the investigated thin-film structure [66].

The variety of optical phenomena, observed during light propagation in optical waveguides, and in devices based on such phenomena, force us to consider only the basic aspects of light propagation in thin-film structures. Some examples of elements and devices necessary for understanding the results are discussed in the following sections. For this reason the list of

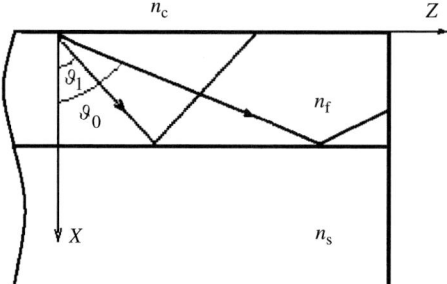

Fig. 2.1. Schematic representation of planar waveguide.

references contains only the papers used for justification or illustration of observed phenomena. More detailed information on these subjects can be found in Refs. [66–72].

2.1.1. Reflection of the Plane Wave from Interface of Two Media

The waveguide propagation of light in thin-film structures can be explained using the approximation of plane waves and by describing a light wave field as the sum of plane waves. Such a simplified approach allows one to illustrate the propagation of light in the waveguide.

First, we consider the process of light reflection from the interface of two media with different refractive indices n_1 and n_2 (Figure 2.2).

In this case the XOZ plane is the incidence plane. All possible electromagnetic fields can be represented as a combination of two types of waves. The first type is the transverse electric (TE) wave, where the vector of electric field $\vec{E}$ is normal to the incidence plane and oriented in the plane of the medium interface. The second is the transverse magnetic (TM) wave. In this case the electric field vector is situated in the incidence plane. Moreover, at the propagation of TE wave in the direction of the OZ axis the vector $\vec{E}$ does not have the longitudinal component ($E_z = 0$). Describing the propagation of the light at angle φ (see Figure 2.2), we obtain the well-known Fresnel formulas [73]

$$r_{\text{TE}} = \frac{\cos\varphi - n_{21}\cos\vartheta}{\cos\varphi + n_{21}\cos\vartheta}, \qquad (2.1.1)$$

$$r_{\text{TM}} = \frac{\cos\vartheta - n_{21}\cos\varphi}{\cos\vartheta + n_{21}\cos\varphi}, \qquad (2.1.2)$$

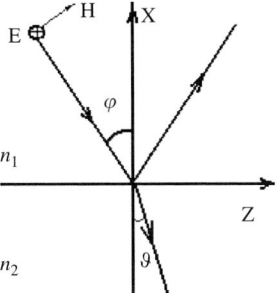

Fig. 2.2. Reflection of the plane TE-wave from an interface.

where $|r|^2 = R$, with R being the energy reflection coefficient, and $n_{21} = n_2/n_1$.

In the case of the reflection of wave from the medium with lower optical density ($n_2 < n_1$) at the incidence angle, which is greater than some angle φ_c satisfying the condition $\sin \varphi_c = n_{21}$, the equality $|r| = 1$ is always satisfacted. This phenomenon is referred to as total internal reflection [74].

2.1.2. Modes in Planar Waveguiding Structures

From the viewpoint of ray optics, the waveguide propagation of radiation in the dielectric layer can be interpreted with the help of total internal reflection phenomenon as mentioned above. In such an approach the light propagation in the direction of the OZ axis (see Figure 2.1) is regarded as the plane wave propagated along the longitudinal axis due to repeated total reflection at the media interfaces. The angle of reflection Θ_m is different for each plane wave and this leads to a difference in the phase velocity [69]. The propagation of plane waves can be described in terms of propagation constants [71]

$$f = a \exp(iux + ihz - i\omega t),$$

$$g = b \exp(-iux + ihz - i\omega t),$$

where $h = k_0 n \sin \alpha$ is a longitudinal propagation constant, $u = k_0 n \cos \alpha$ is the transverse propagation constant, with $\alpha = (\pi/2 - \theta_m)$.

In the case of total reflection, amplitudes of the incident and reflected waves should satisfy the following relationships:

$$g = f \exp(-2i\delta_1),$$

$$f = g\exp(-2i\delta_2),$$

where δ_1 and δ_2 are phase jumps at the upper and lower interfaces, respectively,

$$\delta_1 = \arctan\sqrt{\frac{(h/k_0)^2 - n_a^2}{n^2 - (h/k_0)^2}},$$

$$\delta_2 = \arctan\sqrt{\frac{(h/k_0)^2 - n_s^2}{n^2 - (h/k_0)^2}}.$$

The phase relations mentioned above are satisfied at the condition

$$k_0 n d \cos\alpha - \delta = m\pi.$$

Therefore, the light propagation in thin films takes place only for a discrete series of incidence angles, being greater than the critical angle of total reflection, although the total reflection is observed at any angle $\varphi > \varphi_c$ (see Figure 2.2).

In reality, it is more complicated, and the electromagnetic waves propagate in thin-film structures as optical modes. In this case, the guided mode has spatial distribution of the optical energy in one or two dimensions. The light propagates in the structure as an electromagnetic field, which is described mathematically by the solution of the wave equation. This equation should satisfy the boundary conditions at the interface of different media. The beam patterns depicted in Figure 2.1 correspond to two different modes, TE_0 and TE_1, for example. A certain mode propagates with the phase velocity, that is different for each mode [69], and the distribution of the guided mode field is basically concentrated inside the waveguiding layer (Figure 2.3). All modes have evanescent tails in the surroundings and in the substrate, and contain a harmonic standing wave between them (curves 1 and 2). The evanescent tail is an exponentially decaying field and the

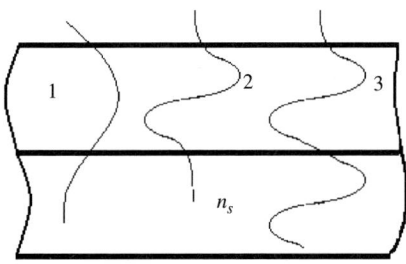

Fig. 2.3. Distribution of the mode field across the waveguide. curves 1 and 2: fundamental and first guided modes, respectively; curve 3: a leaky mode.

harmonic wave is a sinusoid whose frequency varies depending on the mode number [67]. The simplest and the most clear explanation of radiation output is based on the phenomenon of light tunneling into the surrounding medium at total internal reflection.

The depth of the energy penetration into another medium is

$$l = \frac{d\delta}{dn_f} \frac{\sqrt{n_a^2 - n^2}}{2\pi n}, \qquad (2.1.3)$$

where δ is the phase jump at total internal reflection, and n_a and n_s are the refractive indices of the surrounding medium and the substrate, respectively [72]. This effect will be useful in the further analysis of the surrounding properties using waveguide techniques.

2.2. Dispersion Curves and the "Cut-Off" Condition

The characteristics or dispersion equations for TE and TM modes can be found from Maxwell equations and boundary conditions [71]:

$$k_0 d \sqrt{n^2 - (h/k_0)^2} = \arctan \sqrt{\frac{(h/k_0)^2 - n_a^2}{n^2 - (h/k_0)^2}}$$
$$+ \arctan \sqrt{\frac{(h/k_0)^2 - n_s^2}{n^2 - (h/k_0)^2}} + m\pi, \qquad (2.2.1)$$

$$k_0 d \sqrt{n^2 - (h/k_0)^2} = \arctan \frac{n^2}{n_a^2} \sqrt{\frac{(h/k_0)^2 - n_a^2}{n^2 - (h/k_0)^2}}$$
$$+ \arctan \frac{n^2}{n_s^2} \sqrt{\frac{(h/k_0)^2 - n_s^2}{n^2 - (h/k_0)^2}} + m\pi, \qquad (2.2.2)$$

where $k_0 = 2\pi/\lambda$, λ is the radiation wavelength, $h = k_0 n \sin \vartheta_m$ is the real part of the mode propagation constant, m is the mode number, and d the thickness of waveguide.

It should be noted that the approach of ray optics allows one to obtain exactly the same equations with the help of the expressions given in Section 2.1.2. The graphical interpretation of these equation solutions for the ZnSe film with $n = 2.40$ deposited on the quartz glass substrate ($n_s = 1.45671$ at the radiation wavelength $\lambda = 633$ nm) is depicted in Figure 2.4.

As evident from the figure, all guided modes have a definite critical thickness of film, i.e. the waveguide effect can be observed when the film

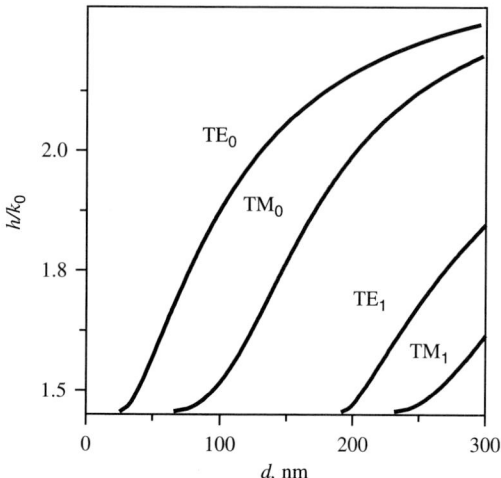

Fig. 2.4. Dispersion curves of the ZnSe waveguide film ($n = 2.40$, $k = 4 \times 10^{-4}$, $\lambda = 633\,\text{nm}$).

thickness is greater than some minimum value d_{min} (which is referred to as the "cut-off" condition). If the guiding layer thickness is less than d_{min} the light localization phenomen in the waveguide is absent. The value of d_{min} differs for each mode, and is defined by the condition of total internal reflection at the interface of the guiding layer and the substrate. In this case $h = k_0 n_s$, and it is possible to derive the following relationships from (2.2.1) and (2.2.2):

$$d_{\text{min}}^{(\text{TE})} = \frac{1}{k_0\sqrt{n^2 - n_s^2}} \left(\arctan \sqrt{\frac{n_s^2 - n_a^2}{n^2 - n_s^2}} + m\pi \right), \quad (2.2.3)$$

$$d_{\text{min}}^{(\text{TM})} = \frac{1}{k_0\sqrt{n^2 - n_s^2}} \left(\arctan \frac{n^2}{n_a^2} \sqrt{\frac{n_s^2 - n_a^2}{n^2 - n_s^2}} + m\pi \right). \quad (2.2.4)$$

It follows from expressions (2.2.3) and (2.2.4) that the minimal layer thickness, required for waveguide propagation of light in the layer, depends on the difference between refractive indices of the substrate and the surrounding. It is evident from the analysis of expressions (2.2.1), (2.2.2) and Figure 2.4 that the $h(d)$ functions asymptotically tend to the value determined by the expression $h = k_0 n$, with increase in the layer thickness.

2.3. Input of Radiation into Waveguide by the Prism Coupler

For the excitation of guided modes, directed by the film, the experimental techniques that use a prism coupler or a prism device of a tunnel excitation of modes are employed. This device provides the selective excitation of a certain mode. The mode excitation is provided by the matching of the light phase velocity in the direction of the *OZ* axis as for the waveguide, and for the incident light beam (see Figure 2.1). This condition of the phase synchronism can be easily satisfied with the help of the prism coupler (Figure 2.5):

$$h_m = k_0 n_p \sin \gamma_m, \quad (2.3.1)$$

where γ_m is the incidence angle of radiation on the prism base and n_p the refractive index of the prism. The similar scheme depicted in Ref. [70].

The light beam with radius a_w is directed onto the prism base. The $n_p > n_g$ condition is satisfied for this prism. Under conditions of the total internal reflection at the prism base, strong coupling of light into the waveguide can occur via resonant-frustrated total reflection, i.e. via evanescent waves in the air gap between the prism and the waveguide. It is evident that $\gamma_m > \varphi_c = \arcsin(n_s/n_p)$. But in the direction of the *OZ* axis this wave

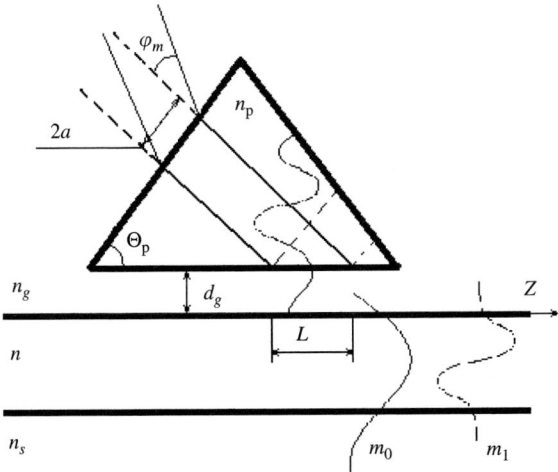

Fig. 2.5. Scheme of the prism coupler: n_p, n, n_s and n_g are the refractive indices of prism, film, substrate and the gap between prism and waveguide, respectively, L and d_g are the interaction length and the gap thickness, respectively.

propagates with a phase constant h. If the thin-film structure can support guided modes then their propagation constants will be h_m. At each internal reflection in the waveguide the interference between the incident and reflected internal beam creates a non-propagating standing wave, which is normal to the reflecting surface. The energy associated with this wave tails out into the surroundings.

If the gap is so small that "tails" of the guided and prism modes overlap, and the φ angle is chosen such that $k_0 n_p = h$, then the energy of the prism mode is transmitted into the guided mode. This condition is defined by the expression

$$h = k_0 n_p \sin\left(\theta_p - \arcsin\frac{n_g \sin \varphi_m}{n_p}\right), \qquad (2.3.2)$$

where θ_p is an angle at the prism base (see Figure 2.5).

By varying the angle φ_m, one can excite several different modes using one prism. The process of energy transmitting into the guided mode is called optical tunneling by analogy with the tunneling effect in quantum mechanics. The length L, where the interaction between prism modes and waveguide takes place, is determined by the prism sizes. The total energy exchange between the prism and waveguide takes place if $pL = \pi/2$, where p is the coupling coefficient, and depends on the gap thickness d_g between the prism and waveguide and n_p, n, n_g, determining the shape of mode "tails". From Figure 2.5, thereby

$$L = 2a_w/\cos \varphi_m = \pi/2p.$$

If a_w is known, one can determine the coupling coefficient. For the Gaussian beam with $I = I_0 \exp(-x^2/a_w^2)$ the coupling efficiency achieves values of ~80% [71].

The prism coupler is often used for the study of optical waveguide properties, and it is not required that $n_p > n$. To excite the guided modes only the $n_p > h/k_0$ condition needs to be fulfilled (see expression (2.3.2)). Usually, one has to ensure that the mechanical pressure applied to the prism is the same in all measurements. This provides constant thickness of gap and the coupling coefficient [66–72, 75]. However, these parameters are not required to be constant (as it will be shown in the next section) because the technique of the thin-film parameter measurement, considered below, takes into account the influence of the prism coupler. It makes possible the determination of the mode propagation constant of the "free" waveguide ($d_g \to \infty$).

2.4. Excitation of Guided Light Modes and Measurement of Their Parameters

In practice, the thin films, obtained by deposition of glasses, oxides, nitrides and semiconductor materials on the glass or semiconductor substrates, are usually multimode waveguides. While investigating such films with application of the prism coupler, which allows one to realize the selective input of light into the certain mode, one can find the incident angle required for the excitation of the separate mode from expression (2.3.2). There are two ways guided mode excitation. The first is the measuring the incident angle of a collimated beam on the entry face of the prism coupler. The excitation angles are measured relative normal to the prism face [27,76] (Figure 2.6a).

The second way is the excitation of a waveguide by the converging beam (Figure 2.6b). As the light is not collimated and enters into the waveguide under different angles, then some part of energy is injected into some guided modes. The radiation of every mode leaves the prism at certain angle corresponding to the given mode. This angle can also be measured [77]. As the waveguide thickness is much less than its width, the light of every mode appears against a background of reflected light in the form of dark lines, and form series of the so-called m-lines [71]. By measuring the angle φ_m, one can determine the mode propagation constants [68] according to expression (2.3.2). This method was quite well studied in theoretical works [77–79], and the problem of experimental equipment was also successfully solved [76, 80, 81]. One can use a gas or semiconductor laser as a source of non-coherent radiation (mercury lamp, for example). In combination with a monochromator it is possible to measure the dispersion of guided modes and to determine the film refractive index at different wavelengths. As evident from expression (2.3.2), the determination error δh of the mode propagation

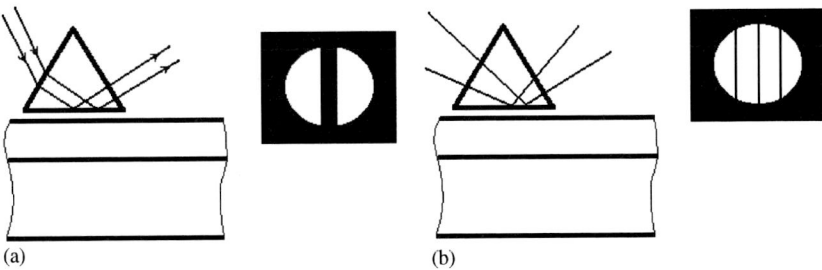

Fig. 2.6. Scheme of the prism excitation of guided modes and mode spectrum observed in the reflected light.

constant is determined by the measurement of errors of the prism angle, the resonant angle of the guided mode excitation and the prism's refractive index. The refractive index is usually determined with the accuracy of 1×10^{-5} [82], and the absolute error of the angle determination is $\sim 2 \times 10^{-5}$. One more source of the measurement errors is the influence of the prism on the measured value of the propagation constant. The dispersion equations are usually obtained by assuming that the media surrounding the waveguide are semiinfinite, but the thickness of the gap between the film and the prism coupler is finite. If the gap thickness is larger than the penetration depth l_1 of the light propagating inside the waveguide, then the gap thickness can be taken as infinitely large. According to Ref. [83], l_1 is equal to

$$l_1 = [k_0(n_w^2 - n_s^2)^{1/2}(n_w^s n^{-s} + n_w^s n_g^{-s} - 1)]^{-1}, \qquad (2.4.1)$$

where s = 0 and 2 for TE and TM waves, respectively, and $n_w = hk_0^{-1}$.

On the other hand, the guided mode excitation is possible only when the depth l_2 of the field penetration from the prism due to the tunnel effect will exceed the gap thickness:

$$l_2 = k_0^{-1}(\sin^2\theta - n_g^2/n_s^2)^{-1/2}, \qquad (2.4.2)$$

where θ is the incident angle of radiation on the interface. As the light intensity is reduced e times at the depth l_2, expressions (2.4.1) and (2.4.2) can be satisfied simultaneously.

Analysis of dispersion equations demonstrates that the value of propagation constant depends on the mode number, polarization of the propagated light, light wavelength and the refractive index of the surroundings [84, 85]. So it is possible to determine the refractive index and the thickness of the waveguide film by measuring the h by varying one of the parameters mentioned above. Thus, there are several techniques of determination of waveguide characteristics.

The dual-mode technique is based on measurements of h_m for modes with different numbers [25]. The application of this approach to thin-film waveguides provides the best accuracy of determining n and d ($\delta n = 2 \times 10^{-4}$, $\delta d = 2$–3%). It is also possible to determine the substrate refractive index for the film guiding more than two modes. The dual-frequency method is the measurement of the mode propagation constant at different wavelengths [86]. While calculating the parameters one assumes that the dispersion of the film and substrate material is the same. It is justified for guiding structures with small difference in the film and substrate refractive indices. In order to decrease the error in the waveguide parameter determination one should measure, if possible, at close wavelengths of the probe radiation. But the measurement error in the dual-frequency method is greater than for the dual-mode one [25]. This technique is unacceptable for waveguides obtained by deposition of an extrinsic material film on the substrate. But if one uses this method in combination with dual-mode technique, the film material dispersion can be determined [87].

The method based on measurement of the mode propagation constants with different polarization is used rarely [88]. Considering the fact that the thin-film structures obtained are practically always anisotropic [89], this method should be used very carefully.

The immersion method is based on the measurement of the mode propagation constant at different refractive indices of the gap [90]. The effect of the h_m value change on the variation of the refractive index n_c of surrounding is widely known and was studied theoretically [84], where the dispersion equations considering the influence of surroundings are given. The analysis of these equations shows that the h_m value increases with the growth of n_c. The influence of the surroundings is more significant for thin-film structures with the high value of $\Delta n = n - n_s$.

2.5. Optical Losses in Waveguides

Up to now we have considered the non-absorbing media and have taken into account only the real part of the mode propagation constant. But real media always absorb the light and are characterized by complex refractive index $\mathbf{n} = n' + ik$, so the propagation constant is also complex, i.e. $h = h' + ih''$ ($h' = \operatorname{Re} h$, $h'' = \operatorname{Im} h$). As the light propagates along the waveguide, there is attenuation of light because of loss of a part of energy to the scattering and absorption centers or because of irradiation into surroundings. The radiation losses usually occur in curved waveguides and hence we will exclude them from our consideration. An exponential absorption coefficient is usually used for the quantitative description of optical losses in waveguides. In this case, the intensity at any point along the waveguide is described by the expression [70]

$$I(z) = I_0 e^{-\alpha z}, \qquad (2.5.1)$$

where I_0 is the initial intensity (at $z = 0$).

The losses (in dB/cm) are related to the α attenuation constant by the relationship $4.3\,\mathrm{dB/cm} = 1\,\mathrm{cm}^{-1}$ and to the imaginary part of the mode propagation constant by $\alpha = 2h''$. The unit "decibel" (dB) introduced here is the measure of energy or power ratio, and is used in communication, electrotechnics, acoustics and so on. This value characterizes difference between levels of energy or power. The number of dB that corresponds to the ratio between two energy or power levels P_1 and P_2 is defined by the expression $10\lg(P_1/P_2)$. When amplitude characteristics are considered (current, voltage and amplitude of propagating light wave) in dB scale, their ratio can be expressed as $20\lg(A_1/A_2)$.

2.5.1. Dispersion Equations for the Imaginary Part of the Mode Propagation Constant

The value h'' determines the attenuation of eigenmodes of the waveguide. The dispersion equations for the real part of the propagation constant of the absorbing waveguide is analogous to (2.2.1) and (2.2.2). The imaginary part h'' of the propagation constant [70,71] for modes TE and TM satisfy the following dispersion equations:

$$h''_{TM} = \left\{ \left[A_1(2n_B^2 - n^2) + 2l\frac{2\pi}{\lambda_0}C_1 \right] \frac{na}{\sqrt{n^2 - n_B^2}} + B_1[2n^2 \right.$$

$$\times (n_B^2 - n'^2) + n^2 n'^2]\frac{n'\alpha'}{\sqrt{n_B^2 - n'^2}} + D_1[2n^2(n_B^2 - n''^2)$$

$$\left. + n^2 n''^2]\frac{n''\alpha''}{\sqrt{n_B^2 - n''^2}} \right\} \left\{ \left(A_1 n^2 + 2l\frac{2\pi}{\lambda_0}C_1 \right) \frac{n_B}{\sqrt{n^2 - n_B^2}} \right.$$

$$\left. + B_1 n^2 n'^2 \frac{n_B^2}{\sqrt{n_B^2 - n'^2}} + D_1 n^2 n''^2 \frac{n_B^2}{\sqrt{n_B^2 - n''^2}} \right\}^{-1}, \quad (2.5.2)$$

$$h''_{TE} = \frac{A\dfrac{n'a'}{\sqrt{n_B^2 - n'^2}} + B\dfrac{n'a'}{\sqrt{n_B^2 - n'^2}} + C\dfrac{na}{\sqrt{n^2 - n_B^2}}}{A\dfrac{n_B}{\sqrt{n_B^2 - n'^2}} + B\dfrac{n_B}{\sqrt{n_B^2 - n'^2}} + C\dfrac{n_B}{\sqrt{n^2 - n_B^2}}}, \quad (2.5.3)$$

where

$$\alpha'' = k_0 k'', \quad \alpha' = k_0 k', \quad \alpha = k_0 k, \quad A = (n^2 - n'^2)\sqrt{n^2 - n_B^2},$$

$$B = (n^2 - n''^2)\sqrt{n^2 - n_B^2},$$

$$C = (\sqrt{n_B^2 - n'^2} + \sqrt{n_B^2 - n''^2})$$
$$\times [(n^2 - n_B^2) + \sqrt{(n_B^2 - n''^2)(n_B^2 - n'^2)}] + 2l\frac{2\pi}{\lambda_0}(n^2 - n'^2)(n^2 - n''^2),$$

$$A_1 = [n'^2 n''^2(n^2 - n_B^2) + n^4\sqrt{(n_B^2 - n''^2)(n_B^2 - n'^2)}]$$
$$\times [n''^2\sqrt{n_B^2 - n'^2} + n'^2\sqrt{n_B^2 - n''^2}]$$

$$B_1 = [n''^4(n^2 - n_B^2) + n'^4(n_B^2 - n''^2)]\sqrt{n^2 - n_B^2},$$

$$C_1 = [n''^4(n^2 - n_B^2) + n''^4(n_B^2 - n''^2)][n^4(n_B^2 - n'^2) + n'^4(n^2 - n_B^2)],$$

$$D_1 = [n'^4(n^2 - n_B^2) + n^4(n_B^2 - n'^2)]\sqrt{n^2 - n_B^2}, \quad n_B = h'/k_0,$$

and n'' and n' are the complex refractive indices of the substrate and the surrounding, respectively.

As we can see, these expressions are quite complicated, but their solution are easy to obtain using computer techniques. The example of curves obtained using expressions (2.5.2)–(2.5.3) is depicted in Figure 2.7. It is

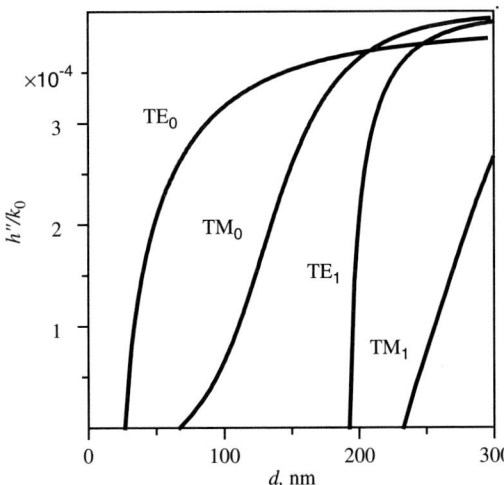

Fig. 2.7. Dependencies of the imaginary part of the mode propagation constant on the film thickness for the ZnSe film ($n = 2.40$, $k = 4 \times 10^{-4}$, $\lambda = 633$ nm).

evident from the figure that h'' asymptotically tends to the value of the absorption coefficient of the layer material. When the film thickness decreases to the value $d_{\min}$, the imaginary part of propagation constant approaches the substrate absorption coefficient.

While studying losses caused by scattering, we should distinguish between the volume and surface scattering. The volume scattering is usually caused by an imperfection of the film structure like impurity atoms, lattice defects, etc. But all these defects and imperfections are negligible in comparison to the wavelength of propagating radiation, and their number is so small that optical losses in waveguides caused by the volume scattering are usually negligible. The propagation losses caused by the surface scattering can be significant, especially for modes of higher order as the wave propagating along the waveguide strongly interacts with its surface. The origin of such losses can be easily understood if we use the approach of ray optics (see Figure 2.1).

The light propagating through waveguide experiences a large number of reflections. For modes of the highest order there can be thousands of such reflections per centimeter of waveguide length. As the surface is not ideally smooth, there are losses due to scattering at every reflection. It can be shown that losses caused by the surface scattering are proportional to the squared ratio of the surface roughness magnitude to the radiation wavelength. It turned out that the measured losses for the waveguide fabricated from tantalum oxide are in good agreement with the data of theoretical evaluations $\alpha = 0.3\,\text{cm}^{-1}$ for the fundamental mode and $\alpha = 2.8\,\text{cm}^{-1}$ for the mode with $m = 3$ [70]. So the scattering losses can be a significant part in total losses for waveguide dielectric films and can achieve values of some dB/cm at the surface roughness equal to $0.1\,\mu\text{m}$. But at the fabrication of thin films with the use of modern technologies the variations of film's thickness are usually controlled at the level of $0.01\,\mu\text{m}$. Hence, the role of losses caused by the scattering becomes less significant.

In semiconductor waveguides, the losses caused by light absorption are usually quite significant and scattering is not so important. The absorption losses are related to the absorption near the band edge and to the absorption on the impurity or surface state levels. Such losses in waveguide are $\sim 10\text{--}200\,\text{dB/cm}$ (or $2\text{--}50\,\text{cm}^{-1}$) approximately. In case of film thickness less than $1\,\mu\text{m}$ it is quite difficult to measure such great attenuation of light by non-waveguide methods.

2.5.2. Measurement of Losses in Waveguides

The techniques of determination of losses in waveguides are based on the input of the definite light power into this film and following the measurement of power at the output of the waveguide. The waveguide type, kind and value of preferential losses define a certain measurement technique. One of the simplest and most precise methods is based on focusing the light on a polished face of the waveguide and further measurement of a transferred total power. These measurements are usually performed at different waveguide lengths by shorting the waveguide [70]. If we construct the dependence of the relative transmission in semi-logarithmic scale it will be a straight line (Figure 2.8).

The optical losses in the waveguide can be determined by the slope of this dependence:

$$\tan \beta = \ln(P_1/P_2)/(l_2 - l_1), \tag{2.5.4}$$

where P_1 and P_2 are the radiation power at the output of waveguide with length l_2 and l_1 ($l_2 > l_1$), respectively.

This destructive method can be used for single-mode waveguides only.

The prism coupler is used to determine the losses in multimode waveguides [80]. The prism position at the light input into the waveguide is usually fixed, and the second output prism is moved after each measurement (Figure 2.9).

The measurement results are processed in the same way as in Refs. [29,30]. The prism-coupling technique has only one essential drawback, that is the strong dependence of coupling efficiency on an amount of clamping at the prism–waveguide contact place [92]. This leads to the measurement error exceeding optical losses.

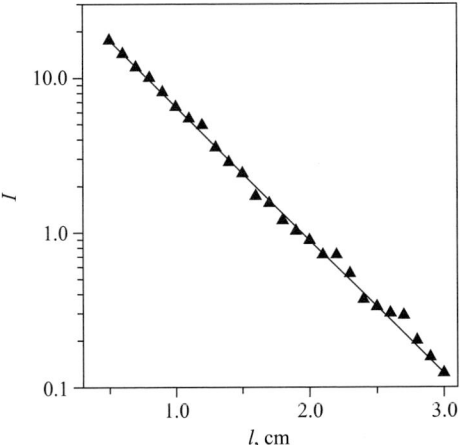

Fig. 2.8. Dependence of radiation power at the output of the waveguide on its length.

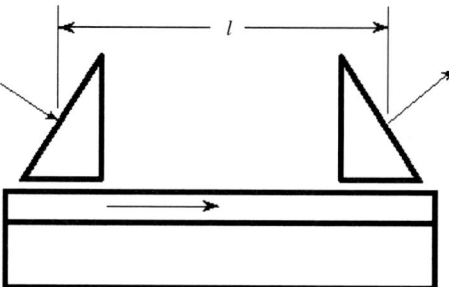

Fig. 2.9. Setup for the measurements of the optical losses in the waveguide by the prism coupling technique.

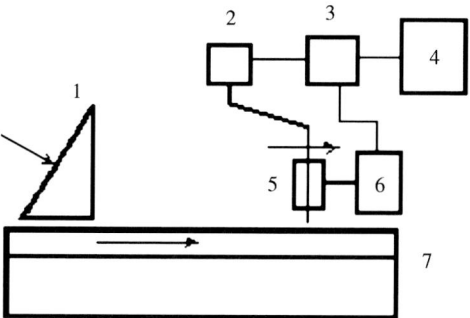

Fig. 2.10. Setup for measurements of the optical losses by scanning the fiber along the waveguide: prism coupler (1), photodetector (2), (3), intensity measuring device (4), fiber (5) scanned by stepmotor (6) along waveguide (7).

As the light propagating inside the waveguide is scattered on inhomogeneities also, there is another possibility to measure the optical losses. The scattered light is propagated in the direction that differs from the direction of the injected radiation and can be recorded (if optical losses are greater than $0.2\,\text{cm}^{-1}$). The photodiode coupled with the optical fiber, collecting the scattered light, is used as a recording device [93]. The fiber is scanned in the direction of the light propagation, so that the distribution of the scattered light along the waveguide can be recorded (Figure 2.10).

The losses per length unit can be determined by the slope angle of line, which is constructed similar to Figure 2.8. Here it should be noted that this technique should be carefully applied for determining optical losses in multimode gradient waveguides because in such layers a coupling between guided modes occurs.

2.6. Leaky and Plasmon Modes in Thin-Film Structures

As mentioned above the modes directed by dielectric thin-film structures can be guided as the leaky or irradiating ones. The conditions of leaky mode existence can be analyzed on the basis of the wave equation solution [94]

$$\nabla^2 E(x,y) + (k_0^2 n_f^2 - h'^2) E(x,y) = 0. \qquad (2.6.1)$$

If we will search the equation solution as $E = E(x,y)\exp[-i(\varpi t - h'z)]$ then in the range of $0 < x < d$ and under the condition $k_0 n_c < k_0 n_s < h' < k_0 n$, its solution is a harmonic function when $E^{-1}\nabla^2 E < 0$, and the field exponentially decays outside the guided layer when $(E^{-1}\nabla^2 E > 0)$ [95]. When $k_0 n_c > h' > k_0 n$ and $k_0 n_c < h'$ the field exponentially decays outside the film and oscillates (has harmonic behavior) inside the film and substrate (see Figure 2.3). In such cases we have modes that leak into the substrate. These modes as guided ones are also characterized by the propagation constant h. So if the film is thick enough, the leaky modes are localized inside the film and propagate along the structure to a great distance. By measuring the propagation constants of leaky modes it is possible to determine the refractive index and the film thickness of such structure as SiO_2–Si, for example [37].

The surface electromagnetic waves or plasmon modes are used for the study of the metallic films and surface layers of bulk metals. Such a mode is the bound state of inhomogeneous TM wave with surface plasmons (or wave of density of free electric charge carriers at the metal surface). The range of existence of plasmons at the interface of the two media with $\varepsilon_1 = \varepsilon_1' + i\varepsilon_2''$ and $\varepsilon_2 = \varepsilon_2' + \varepsilon_2''$ is depicted in Figure 2.11. The peculiarities of the ε dependence on frequency are considered in Section 1.2. The plasmon mode field intensity is exponentially attenuated along the normal to the interface of the two media and in the direction of the light propagation (Figure 2.12). As follows from Figure 2.11, the condition $|\varepsilon_2| > |\varepsilon_1|$ (i.e. in the air $\varepsilon_2 < -1$) is necessary for the existence of the plasmon mode [96]. The plasmon mode propagation constant is related to the dielectric permittivity of the medium by the expression

$$h = h' + ih'' = k_0(\varepsilon_1 \varepsilon_2/(\varepsilon_1 + \varepsilon_2))^{1/2}. \qquad (2.6.2)$$

The phase velocity of plasmon modes is less than that of the volume electromagnetic waves and hence it is impossible to convert a volume wave into the surface wave and vice versa. The special techniques used for exciting the plasmon mode include the prism-coupling technique mentioned above [97,98]. The plasmon propagation length is determined by the imaginary part of the propagation constant,

$$L = 1/2|h''|. \qquad (2.6.3)$$

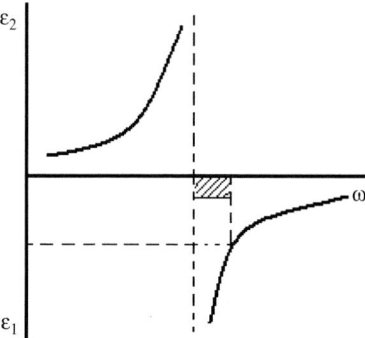

Fig. 2.11. Plasmon mode existence area (crosshatched) [96].

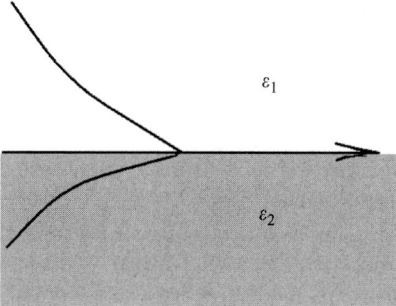

Fig. 2.12. Field distribution of the plasmon mode propagating along the interface.

The field of electromagnetic wave penetrates into the metal to the depth l_p, which is defined as the distance where the wave amplitude becomes e times smaller:

$$l_p = 1/(\mathrm{Re}(k_0\varepsilon_2 - h^2))^{1/2}. \tag{2.6.4}$$

In the case of the prism excitation of plasmon modes, application of recording of the resonant minimum in angular distribution of the light beam reflection coefficient allows one to measure the refractive index and the absorption coefficient of bulk metals or films with the thickness $d > l_p$, the accuracy of such measurement being of 5–10%. The thickness and the refractive index of thin films ($d < l_p$) can be also determined by using this technique [99,100].

In general terms we have considered all types of optical modes directed by thin-film structures. As it will be shown in the following sections, we will try to use the characteristics of these modes in order to determine spectral-optical parameters and thickness of metals, semiconductors and dielectric thin films.

2.7. Fabrication of Waveguiding Structures

Before starting to investigate thin-film structures let us consider the methods of their fabrication. The different methods of the deposition of the film made from different materials are used in optics and microelectronics [101–103]. The choice of the technique is caused by the investigation objectives and the available equipment. One of the first methods used in integrated optics for fabrication of optical waveguides is the traditional technique of thermal evaporation in vacuum. The waveguides obtained by this method have significant losses (equal to 10–20 dB/cm) [104] caused by the foreign matter atom impurity, which causes scattering and absorption of the light. A great number of fabrication methods based on solid sputtering allows one to obtain the waveguides with different functionalities in the form of films are as layers with non uniform distribution of the refractive index over cross-section [106]. The methods of electron-beam evaporation [107], RF sputtering [107–111] and reactive-cathode sputtering [112,113] are widely used for fabrication of dielectric covers as planar waveguides. The high-quality waveguides fabricated from the silicon oxy-nitride films are obtained by MCVD [114]. By using the sputtering techniques it is possible to obtain crystalline layers as layers of binary semiconductor compounds [116–118].

The gradient waveguides are obtained in crystalline and amorphous materials by the effect that increases the refractive index of the surface layer of the bulk sample and makes possible the waveguide propagation of the light in such a structure [22]. Some methods use the implantation of impurity atoms, which substitute separate atoms in the bulk material lattice, hence the increase of the refractive index takes place. These include methods such as the ion implantation [119], solid-state diffusion [120] and ion exchange [121]. The semiconductor technology provides one more possibility to fabricate the waveguide. The existence of free carriers in such materials decreases the refractive index in comparison to that of the carrier-depleted material. So, if the charge carriers were removed from the surface layer of the semiconductor substrate, the refractive index of this area would be greater than one of the substrate. In such situation the light will be guided by the semiconductor structure. The proton irradiation is one of the methods of carrier concentration decrease [122]. In this case, the variation of the refractive index Δn, caused by the carrier concentration variation $\Delta N = N_s - N_f$, is

$$\Delta n = \Delta N e^2 / 2 N_s \varepsilon_o m^* \omega^2, \qquad (2.7.1)$$

where ω is the radiation frequency, and m^* and N_s are the carrier effective mass and the initial concentration of carriers, respectively [70].

The epitaxial growth technique of fabrication of the guiding structure is universal for crystalline materials. It is caused by the fact that the chemistry of a grown layer can be varied in order to obtain the thin-film structure with required properties [123–125].

All techniques of fabrication of the guiding thin-film structure and gradient layers mentioned above allow one to create the structures, which can direct optical modes. Hence, these structures can be studied by the prism-coupling technique.

Chapter 3
New Applications of the *m*-Line Technique for Studying Thin-Film Structures

3.1. Recording the Spatial Distribution of the Light-Beam Intensity. 41
    3.1.1. Photodetectors in Waveguide Measurement Setups. 43
    3.1.2. Light Sources and Characterization of Light Beams 50
3.2. Techniques and Setup for the Measurement of Light Beam Intensity and its Spatial Distribution 61
    3.2.1. Mathematical Processing the Recorded Spatial Distribution 64
    3.2.2. Determination of the Light-Beam Parameters 68
3.3. Spatial Distribution of the Light Beam Intensity Reflected from the Prism Coupler and Measurements of Thin-Film Parameters 70

The prism-coupling techniques used for measuring the parameters of thin films and surface layers are based on recording reflected light-beam intensity in the location of the dark *m*-line excitation of guided modes by a prism. This provides the possibility to determine the real part h' of the mode propagation constant h ($h = h + ih$, $h'' = \text{Re } h$, $h = \text{Im } h$), correspondingly limits the capabilities of these measurement methods to determine the film parameters such as the refractive index and film, thickness. However, the experimental data indicate that the observed picture contains much more information. If we examine closely a separate line it is possible to find that the *m*-line has its own "fine" structure (Figure 3.1).

The intensity distribution of the reflected radiation differs for films with different optical losses. Thus, the measurement of the characteristics of such distribution and to associate them with the thin-film parameters, was considered as an objective.

3.1. Recording the Spatial Distribution of the Light-Beam Intensity

In determining thin-film parameters there is a need to measure the optical radiation and spatial distribution as the absolute value of the intensity (power). In many experiments on light detection the accurate measurement

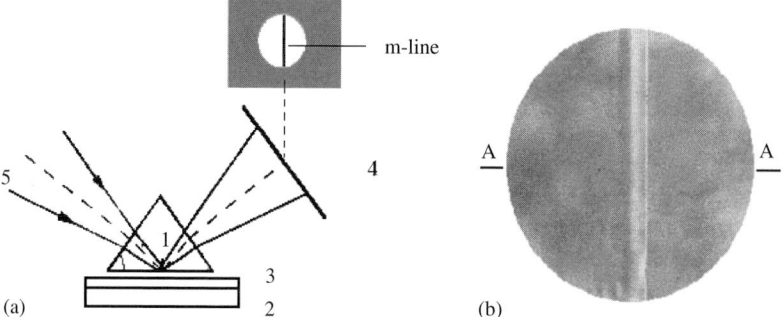

Fig. 3.1. The prism-coupling setup for excitation of the guided mode and "fine" structure of the *m*-line.

of its intensity forms the basis of successful operation [129]. This problem exists in interferometry, Fourier spectroscopy, laser anemometry [130], etc.

While investigating, the spatial distribution of intensity the results of which are discussed below, the low-intensity radiation was used ($\sim 10^{-3}$ W). Hence the technique of direct photoelectric transformation of the light into electric signal with its following processing was applied for measuring light intensity [131]. The measurement setup (Figure 3.2) consisted of matching optical elements, photodetector, and a device for recording and processing the electric signal. In the process of photoelectric recording, the variations of the probe light intensity and the coefficient of its transformation into electric signal, which do not depend on the presence of the investigated sample, caused the change of the measured value:

$$R = \frac{I'}{I} = k_I k_f k_s k_L \frac{U'}{U}, \qquad (3.1.1)$$

where I and I' are readings of the measuring system with and without the investigated sample, respectively, k_I, k_f are the coefficients characterizing changes in the light source and photoelectric transformation parameters, respectively, k_I, k_f are the characteristics of sensitivity and linearity of the photodetector, respectively, and U' and U are the photodetector indications that depend on the probe radiation intensities I' and I, respectively.

The term "intensity" can be referred any energy parameter of radiation: average power, light intensity and energy, etc.

Taking into account the small values of the measured losses in films one can obtain from expression (3.1.1),

$$\Delta R = \frac{2\Delta U}{U} + \sum \frac{\Delta k_i}{k_i}, \qquad (3.1.2)$$

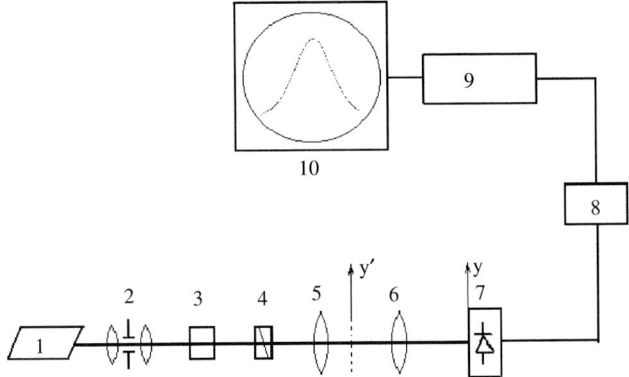

Fig. 3.2. Experimental setup used for measuring the spatial distribution of intensity in cross-section of the light beam: light source (1), collimator (2), attenuator (3), polarizer (4), lens (5, 6), photodetector (7), intensity measuring device (8), analog digital converter (9), computer (10).

where ΔU is the stochastic variation of the signal and $\sum \Delta k_i/k_i$ the total error of the applied measurement technique [3].

As it is evident from the expression (3.1.2) the value of the optical losses, which can be determined by direct measurement, depends on the doubled error of the recorded energy parameter and by the error that is typical for the measurement technique. While recording the reflected radiation, for example, the second component included the error caused by the irregularity of the photodetector sensitivity, beam axis shift, uncontrolled variations of the photodetector sensitivity and of the radiation intensity. In general, the systematic error of the technique can be minimized even to zero. In this case, the value equal to the doubled ratio of the noise to the useful signal determines the lower limit of the measured losses.

As shown below, the results of thin-film parameter measurements by using the techniques considered in this book do not depend on the absolute value of the measured light intensity but are determined by its relative distribution in the recording plane. The advantage of these techniques is that, it does not require the measurement of the absolute value of the light intensity.

3.1.1. Photodetectors in Waveguide Measurement Setups

In optoelectronic recording systems the photodetector is the main part of the measuring system. The detectors used in the measuring systems differ by their construction, functionality and physical mechanisms, defining the operation of the photodetectors. The techniques and devices of the laser photometry are specified in detail in Ref. [133]. The photoelectrical schemes of direct detection record the signal with the help of the photoemissive or photoconductive devices [134–139]. We will not consider thermal detectors, which are usually used in the infrared spectral range. The basic characteristics and principles of operation of CCD-array detectors are considered in Refs. [140–142]. The small size, low-power consumption, high sensitivity, and small response time of semiconductor devices are advantages that define the application of photodiodes as measuring systems.

Depending on the parameters of the recording system and the detector the variations of the recorded signal can be caused by noise of different types [132]. It means that the measurement accuracy is restricted by statistical deviations in the mean value. Among the errors caused by the environment, not related to the measuring device, we can distinguish a partial absence of the photodetector plane in the light. As usual it is concerned with the positioning errors or instability of the laser beam profile. But when measurements are performed by the photodetector with certain coefficient of conversion of the light energy into electrical signal, the results are always underestimated. Another reason for the measurement errors is multiple light reflections between the source and the detector, which can lead to distortion of the amplitude as well as the signal shape. To reduce this error one needs a precise positioning of the working area of the detectors relative to the reflecting interfaces of the source and the optical elements. In the case a laser is used as a coherent light source the formation of fringe pattern is noticeable. For example, it leads to heterogeneous transmission of a thin glass plate. The light polarization can also affect the transmission coefficient of elements such as prisms. The interference and polarization phenomena cause systematic errors in the variables, and hence these parameters should be checked periodically.

If we take into account or remove all the errors caused by the environment, during measurement by the most precise and modern devices the obtained results will still have some spread in values, which have a statistical origin. First of all the random thermal motion of the atomic particles causes the so-called thermal noise in all electric devices. It is concerned with the creation of statistical oscillations of charge density, which leads to the rapidly fluctuating voltage between the edges of the conductor sample – the noise voltage. The noise effective voltage U_R is specified by the Nikewist formula, which is derived from the conditions of thermodynamic balance and the law of energy distribution over the freedom degree [138]:

$$u_R = 4k_B T R \Delta v,$$

where R, T, and Δv are the conductor resistance, temperature, and operating frequency range, respectively.

At the input resistance of the amplifier equal to 1 MΩ and $\Delta v = 100$ MHz, the noise voltage is equal to ~1 mV at room temperature. This value is quite significant at the recording signal equal to 100 mV, for example.

In semiconductor devices there is a specific kind of shot noise, which is called the generation–recombination noise. The rate of generation and recombination of charge carriers determines its value. The effective value of this noise is described by the expression

$$I_{GR} = A \frac{I_0^2}{1 + v^2/v_g^2} \Delta v,$$

where I_0, v and $v_g = 1/2\pi\tau$ are the current in the semiconductor sample, operating frequency and threshold frequency, respectively, and τ is the carrier lifetime.

When $v > v_g$ the value of noise decreases to $1/v^2$. If the discrete origin of charge carriers causes the shot noise, then the quantization of the electromagnetic field leads to the fluctuation on photon number. Therefore during direct detection by an ideal photodetector, the power of quantum noise will be described as [138]

$$P_R = 2\hbar\Delta v,$$

which does not depend on the light power.

Unlike thermal noise, whose level decreases at high frequencies, quantum noise linearly increases with the operating frequency. When $hv/k_B T > 1$, the quantum noise becomes greater than the thermal noise. At room temperature it corresponds to the visible and infrared spectral range. If the measured value is so low that an induced signal passing through the detector, converter and amplifier is less than the noise of the measuring device, direct measurements are not possible. Application of phase-sensitive amplifiers containing RC circuits as frequency filters allows one to limit the frequency band and decrease the noise.

Let us consider measurements of the power density of the light beam. At the output of the measuring device (Figure 3.3), there is a signal with voltage U_0 and current I_0. Let us describe the total value of the noise by the equivalent noise R_{eq} and the load impedance by R_L.

During direct measurements the signal-to-noise ratio is defined as the ratio of appropriate powers:

$$\frac{P_S}{P_N} = \frac{U_0^2}{U^r R_{eq}} = \frac{U_0^2}{4kT R_{eq} \Delta v}.$$

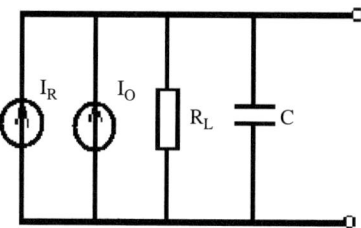

Fig. 3.3. The equivalent scheme of photodetector.

In order to make the thermal noise of load impedance lower than the shot noise it is necessary to satisfy the condition

$$R_L > 2kT/eI_0,$$

i.e., at room temperature and $I_0 = 15\,\text{mA}$ load impedance should be greater than $300\,\Omega$. The large impedance leads to the limitation of the operating frequency band due to the short-circulating effect of capacity. As shown in Ref. [133] in the case of the connection of the photodetector, according to Figure 3.3, the wide spectrum range determining the noise of the detector is reduced to a narrow band:

$$\Delta\gamma_{\text{ef}} = \frac{1}{2\pi R_L C}.$$

The load impedance decreases the signal-to-noise ratio, and therefore the choice of the R_L value is a trade off between the decrease of the thermal noise and the expansion of the operating frequency band.

The sensitivity of photodetectors is its important parameter. The spectral sensitivity is determined by the ratio of the output signal (current i_p or voltage) to the incident light intensity. The sensitivity depends on the wavelength (photon energy) and is related to the quantum efficiency η_p by the expression [139]

$$S = i_p/I = e\eta_p/\hbar\omega.$$

If the wavelength is in microns and the photon energy in electron-volts, then sensitivity is determined as

$$S = e\eta_p\lambda/1239.$$

The quantum efficiency is equal to the ratio of the number of collected carriers to the number of photons incident on the photodiode. In real devices a part of the light power is reflected from the detecting area of p–n junction. Besides, a part of the power absorbed in the space-charge region

(SCR) depends on its width d_{SCR} and the absorption coefficient α for incident radiation. As a result the quantum efficiency is given by

$$\eta_p = (1 - R)(1 - e^{-\alpha d_{SCR}}),$$

where R is the reflection coefficient of the photodiode surface.

To maximize quantum efficiency the SCR should be quite wide for the condition $\alpha d_{SCR} >> 1$ to be satisfied. In case of interband excitation and $\alpha = \sim 10^4$ cm^{-1} the value of d_{SCR} is of some micrometers. For such devices the threshold of sensitivity (or minimal power that can still be recorded), determined on the assumption of the signal-to-noise ratio equal to unity, is about 0.5 µW. There are various methods to determine the detector sensitivity: in a simple method the sensitivity of the photodetector under test is compared with the known sensitivity of the standard detector. In this case, measurement of the incident light intensity is not required as it is similar to that measured by tested and standard detectors. If there is a source of radiation with a certain power the detector sensitivity can be measured by changing the incident light intensity in known proportions. One should take precautions to protect the photodetector from scattered light. A system of apertures or a modulated light beam with further recording by selective amplifiers should be used. Gray glass filters are often used to change the incident light intensity. But we should remember that they can be used while measurements are performed in a narrow spectral band. The application of a spectroscopic cell with a dye solution of certain concentration also helps to change the light intensity. But the nonlinear effects of bleaching or nonlinear reflection may appear and they should be accounted while using the laser radiation.

The techniques described above can be useful while measuring integral sensitivity, sensitivity threshold, linearity, etc. The linearity range of the photodetector is defined as the range of input power, where the power of output signal is proportional to the power of the incident light. In practice, one should test the linearity of the whole measuring device. It is a simple technique that uses gray glass filters, as described above, and requires a preadjustment of these filters with high accuracy. A more accurate technique would be summing of the reaction of the measuring device influenced by some light beams. An example of such summing technique is the recording of the variations in the device reading caused by the additional illumination of the detector against a background of the basic light beam.

The range of spectral sensitivity is another characteristic, which should be taken into account while choosing the photodetector. The light absorption in medium is described by expression (1.2.21), and the absorption coefficient strongly depends on the wavelength of the light. A noticeable photoresponse is recorded only in a certain spectral range. The wavelength, λ_k, limiting this region from long-wave side, is called the photoelectric threshold, and is related to the photon energy, which is equal to the band-gap energy, E_g [150]:

$$\lambda_k = 2\pi\hbar c/E_g.$$

One- and two-element semiconductors (Ge, GaAs, InP, etc.) are used as materials for manufacturing photodetectors. As all semiconductor materials have a certain width of band gap, so the maximum wavelength of the radiation, which the detector can record, also has a fixed value. It is possible to change the value of E_g and to create detectors with selective sensitivity for different spectral ranges by changing a semiconductor compound ($Ga_{1-x}Al_xAs_x$, $In_xGa_{1-x}As_yP_{1-y}$).

The semiconductor detector in combination with an amplifier satisfies all requirements imposed on the measuring devices for low-power radiation measurement. High sensitivity, wide range of linear characteristic, small size, and low control voltages of such devices allow one to solve efficiently the problems of measuring the parameters of optical radiation [143]. The techniques of light-intensity transformation by averaging the light beam cross-section at the wide aperture of the integrating detector (photodiode) are quite simple, and allow one to decrease the influence of spatial oscillations of the light intensity. But the error of such measurements essentially depends on the shape of the light beam. In contrast to the single-channel system of recording, the approach of the discretization of the light beam intensity is more informative and allows one to compensate the imperfection of the optical system. In such a recording system there is the possibility of determining the shape and the orientation of the light beam. A comparative analysis of these techniques, which takes into account the characteristics of the photodetector, shows that the devices using the principles of discretization provide a higher noise immunity of the measuring device [138]. The discrete transformation of the light intensity is based on the recording of the signal at separate points by matrix detectors. Detectors of the CCD-array type are more suitable for these purposes [137,140,141]. High stability of parameters, high accuracy of discretization (no more than 0.3 µm), and a wide dynamic range (up to 60 dB and more) are the advantages of such detectors [138].

The possibility of application of such detectors to record the light-beam intensity distribution while measuring thin-film parameters depends on whether their parameters meet the experimental requirements. A typical angular size of the separate element in the pattern of intensity distribution is about 0.01° of arc. The diameter of the light beam usually varies from 100 to 300 µm. Since the element of CCD array is $\leqslant 10$ µm, these detectors are usually able to provide the necessary sampling step of the recorded intensity distribution. The sensitivity of such detector can be estimated as follows. Let the power of the radiation source in the circuit depicted in Figure 3.2 be 10 μW. The Gaussian light beam undergoing some reflections at the optical element surfaces and being partly absorbed in the film, will have the following intensity at the photodetector input:

$$I = \kappa(1 - R)I_0 \exp(-x^2/a_0^2),$$

where κ is the light attenuation coefficient of the film, R the coefficient of the total reflection of the light beam in optical device, I_0 the intensity at the center of the light beam, a_0 the typical size of the light beam and x the coordinate in the detector plane.

$x \approx 1.4a$ for the area of the recorded pattern, where the intensity is relatively small.

Hence, the estimated power of the radiation incident on one cell at $R = 0.5$, $\kappa \approx 0.9$ is about 3×10^{-8} W. The sensitivity threshold for some photodetectors is equal to 10^{-12} W, which provides fine recording of the light beam intensity distribution in a wide range of spatial frequencies.

While choosing the photodetector one should take into account all the characteristics of the detector mentioned above in order to achieve necessary sensitivity and accuracy of detection. Parameters by some photodetectors are given in Table 3.1. It should be noted that sensitivity of avalanche photodiodes is usually 10 times higher than the values given in Table 3.1. A comparison of the characteristics of the photodetector at wavelengths 1.3 and 1.55 μm can be found in Ref. [135]. We can take into account possible changes in the integral sensitivity and the detection threshold caused by the variations of the light intensity while comparing the incident radiation intensity and energy characteristics of the detector. Not only the useful signal but also the background illumination is usually taken into account. One of the important steps is the coordination of the output aperture of the optical

Table 3.1. Photodetector characteristics

Detector type	Spectral range (μm)	Response time (s)	Sensitivity (A/W)
PhED	0.3–1.2	$<5 \times 10^{-11}$	
Photoresistor	<2–7	$\leqslant 10^{-9}$	
Photodiodes		$\sim 10^{-5}$–10^{-10}	
Si	0.35–1.1		0.2–0.5
Ge	0.4–1.6		0.5–1
GaAs	0.4–0.9		0.2–0.5
InGaAsP	1.0–1.55		0.3–1
CCD-array	0.3–1.0	$\sim 10^{-3}$	6×10^{-12}

scheme with the allowed light angles on the input area, since light absorption depends on the incident angle. If the detector is placed in the focal plane of the lens, then the size of the spot will affect the noise level because of irregularity of the detector sensitivity over the input area. Therefore, it is desirable to use an optical system that provides maximal illumination of the detector input area.

During the recording of low-intensity light beams, attention is paid to the matching of the parameters of the detector and also to other elements of the electronic circuit. For example, the amplifier should not at least decrease the signal-to-noise ratio. Depending on the photodetector type this matching can be realized by utilizing a cathode repeater for high-resistance detectors or a transformer input for low-resistance detectors [131]. Taking into account these and other factors [135], one can accurately determine the parameters of low-intensity light beams.

3.1.2. Light Sources and Characterization of Light Beams

The sources of a non-monochromatic and coherent radiation are used in the measurement setup and devices. Our goal is to investigate the specific characteristics of radiation and light beams, which will define the measurement accuracy and correct interpretation of the obtained results. But the principles and design philosophy of the radiation sources are out of the scope of this chapter. Some of the features of coherent light source functionality are used only for explanation of the observed effects. One can find a more detailed information and fundamental consideration of the operation principles of laser and non-coherent sources in Refs. [12,144–149].

The light-emitting diodes (LEDs), glow lamps, and discharge lamps are used as sources of non-coherent light. Their radiation has a low coherence. The radiation spectrum of such devices has $\Delta\omega$ values in a specific range due to the temporal instability of the amplitude and the phase. The value of $\Delta\omega$ is used as a parameter characterizing light-source monochromaticity. Besides, the irradiated light does not have a regular direction in space, and hence non-coherent radiation has a low orientation. Glow lamps are the first electric light source to obtain a wide application because of simple operation and suitable spectral structure of radiation. The working element of this lamp is usually manufactured in the form of a spiral tungsten wire, suspended in vacuum or in the atmosphere of a rare gas. The main part of radiation is in the infrared spectral region. One can find extensive information about thermal sources of radiation in Ref. [151]. While choosing a lamp one should take into account the operation conditions provided by the manufacturer. But sometimes, especially working in the high-frequency spectral region, one can shorten the lamp's service life by working in the "hard" regime of its power supply in order to get a high intensity of light. It is worth noting that the lamp's power supply regime must be maintained accurately because if the filament voltage changes at 1%, the light intensity variation is approximately equal to 3.5% [151]. While using discharge lamps filled with different gases and metal vapors it is possible to vary the radiation spectrum within wide limits and concentrate the main part of the radiation in the necessary spectral range, depending on the gas filling the lamp [149]. For LEDs fabricated on the basis of multicomponent semiconductors, the changes in the band gap depend on the material compound. This fact allows one to create light-emitting devices with different wavelength of radiation. For example, for a compound like $In_xGa_{1-x}As_yP_{1-y}$ the wavelength of the radiated light changes from 1.0 to 1.6 μm at different x and y values [53]. The typical half-width at half-maximum (HWHM) of the spectral band for LEDs in a visible range are given in Table 3.2.

One of the main characteristics of LEDs is the power of the radiated light. For AlGaAs light-emitting diodes it is up to 100 mW, when the injection current is equal to 100–200 mA. The central wavelength of the spectral band is within the range 730–900 nm [148a], and the spectral width is equal to 30–60 nm. We should note that the radiation divergence angle for such diodes is 120–180°. This makes it difficult to obtain in-focus rays and decreases the effectiveness of the light-emitting diode application in optical devices. The light-emitting diodes based on the InGaAs semiconductors have the radiating power of 1–3 mW, the main wavelength of radiation equal to 1.1–1.5 μm and spectral width of about 15–30 nm. Parameters of commercial LEDs are given in Table 3.3.

In spectroscopy, lasers instead of the usual spectral lamps is used as radiation sources in most cases. In stimulated irradiation, photon emission is caused by the existing electromagnetic field. The probability of the excited power state population is proportional to the spectral density of the energy of the existing radiation field with frequency ω_0. If the light with frequency ω_0, approximately equal to the frequency ω, affects the electron situated in

Table 3.2. HWHM for light-emitting diodes [148]

Semiconductor compound	λ_{max} (nm)	$\Delta\lambda_{0.5}$ (nm)
GaAlAs	650	22
AlGaInP	639	19
AlGaInP	620	17.5
AlGaInP	594	16
InGaN	526	35
InGaN	470	25

Table 3.3. Some parameters of light-emitting diodes [148]

LED type	Luminescence color and λ_{max} (nm)	Frame diameter (mm)	Luminous intensity at current of 20 mA		Radiation angle $2\theta_{0.5}$ (deg)	Maximum current (mA)	Country and firm-manufacturer
			No less	Type			
HLMP-CB30	Blue, 475	5	0.4	0.56	30	30	USA, HP*
NSPB510S	Blue, 470	5	—	0.65	30	30	Japan, Nichia
NSPE510S	Blue-green, 500	5	—	1.7	30	30	Japan, Nichia
HLMP-CE30	Blue-green, 505	5	—	2.1	30	30	USA, HP*
HLMP-CM30	Green, 526	5	1.0	1.75	30	30	USA, HP*
NSPG510S	Green, 525	5	—	2.2	30	30	Japan, Nichia
TLGA159P	Green, 574	5	0.27	0.7	20	50	Japan, Nichia
U-118G	Green, 575	5	0.4	0.6	25	50	Russia, Optel
U-118B	Green, 565	5	0.15	0.3	25	30	Russia, Optel
HLMP-DL25	Yellow, 592	5	1.0	2.8	23	50	USA, HP*
U-118D	Yellow, 592	5	0.7	1.5	25	50	Russia, Optel
U-114D	Yellow, 592	8	1.5	2.0	10	50	Russia, Optel
U-164D	Yellow, 592	10	6.0	10.0	4	50	Russia, Optel
HLMP-BB25	Red, 630	5	1.0	1.8	23	50	USA, HP*
U-118B	Red, 630	5	0.7	1.5	25	50	Russia, Optel
U-114B	Red, 630	8	2.5	3.5	10	50	Russia, Optel
U-164B	Red, 630	10	10.0	20.0	4	50	Russia, Optel
WU-7-750SWS	White	5	—	3.0	20	30	Germany, Wustlich

*Hewlett Packard.

the upper energy level (in gas) or in the conduction band (in semiconductors), the radiation with frequency ω_0 and propagation direction of the incident light appears. Here it should be noted that the stimulated emission (lasing) occurs, if the "inverted population" takes place. The inverted population occurs when there are more electrons in the upper energy state than there are in the lower energy state. For effective use of induced radiation the light-emitting medium is placed into the cavity of the resonator, as shown in Figure 3.4.

Gas quantum generator with an electric excitation system is a prevalent type of laser; the He–Ne laser is a typical example. The radiation wavelength of such laser can be 0.6328, 1.15 and 3.39 μm depending on its embodiment. The laser is usually produced by a Fabry–Perot resonator, where the space between the mirrors is filled with an active medium. According to the laws of electromagnetic theory the electric field in the plane normal to the Z direction (Figure 3.4) is similar to the distribution shown in Figure 3.5. The similar distribution is depicted also in Ref [129]. The mode of zero, first or second order, is defined by the number of m and l curves crossing the X- or

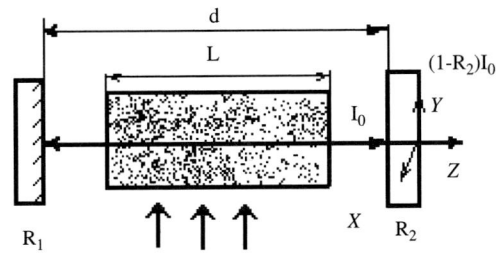

Fig. 3.4.   Scheme of laser structure.

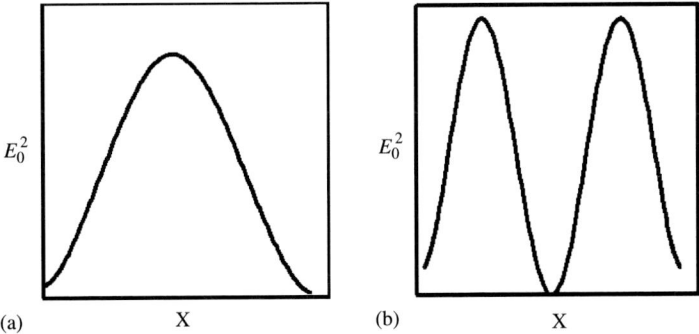

Fig. 3.5.   Mode field distribution in resonator for transverse modes of zero (a) and first (b) order.

Y-axis in the specified plane. The m and l constants are used to denote the order of the transverse mode of the resonator. In the process of successive reflections from the mirrors of the resonator the wave undergoes multiple diffraction. The Fabry–Perot interferometer can be replaced by an equivalent system of an infinite number of diaphragms, where the diffraction takes place. If this system is illuminated by a plane wave, the wave washed out by diffraction will have amplitude distribution at the output similar to Gaussian distribution because of a greatly number of "diaphragms".

It is evident from the consideration of the mode parameters that it is quite a widespread case. In particular, the intensity distribution for the fundamental mode of the laser resonator is Gaussian [129]. The intensity distribution over the cross-section of such light beam is described by the expression

$$I(x,y) = I_0 e^{-r^2/a_0^2} \qquad (3.1.3)$$

where $r^2 = x^2 + y^2$, a_0 is the beam width, which has been defined as the e^{-1} Gaussian power width, and I_0 is the intensity in the beam center.

While choosing a light source of radiation for measuring spatial intensity distribution of the reflected light beam, one should pay attention that the intensity distribution conforms to the distribution of the fundamental mode, which is shown in Figure 3.5. It means that the shape concordance of the light beam intensity distribution to the function, given by expression (3.1.3), is the main criterion for the application of the light source. Such quantity as the beam radius on mirror (often called the spot size) is used to evaluate the beam size. For confocal resonator it is defined by the expression

$$a_s = \sqrt{\lambda d/\pi} [2d/b^1 - (d/b^1)^2]^{-1/4}, \qquad (3.1.4)$$

where $2db^1 = b^2 + d^2$, b is the radius of mirror curvature, and $d = L$.

In this case a_s is the radial distance where the field amplitude is e times lower in comparison to its maximum on the axis.

Semiconductor lasers are also sources of coherent radiation, but in contrast to gas lasers the radiating transitions in semiconductor lasers take place between the energy bands, and the pumping is performed by transmission of the electrical current through the diode. This gives the possibility to generate modulated radiation. To obtain laser generation a multilayer heterostructure is usually created. In this structure electrons are "locked" in the active layer, and both the open faceplates of this heterostructure, which are perpendicular to the axis of light beam, serve as mirrors. This is how a Fabri–Perot resonator works. The resonator length is usually about 300 μm. The average depth of diffusion of electrons, injected into the active layer is equal to 1.2 μm, and the active layer thickness d_a is determined by this value. Besides, if we want to generate only the mode of the lowest order from the set of all possible transverse modes, we have to use even smaller value of $d_a \sim 0.1$ µm. The active layer thickness in the direction normal to the light propagation should also not be greater than this value. These are the conditions for propagation of transverse modes of the lowest order. The laser used is a semiconductor laser with a heterostructure of strip-line type. Because of light scattering by lattice and light absorption by impurities, and because of higher values of the refractive index of active layer (the refractive index for GaAs is equal to ~ 3.5) and, therefore of low reflectance of the mirror ($\sim 30\%$), the losses in the resonator will be considerable and its Q-factor will be small. Usually there are about 100 longitudinal modes within the spectral range of amplification for the diode of such type, and there are practically no differences between the gain of different modes. For this reason the modes are generated simultaneously (Figure 3.6).

The frequency spacing between the adjacent modes is given by the expression [136]

$$\Delta \lambda = \frac{\lambda^2}{2nL(1 - \frac{\lambda}{h}\frac{dn}{d\lambda})}. \qquad (3.1.5)$$

But for improvement laser emitting diodes the picture of generation is shown in Figure 3.7. The spectrum width of the radiation determined is defined by the characteristics of the resonator and the active medium. The active medium is characterized by the spectral band width Δv according to the Geinzberg relationship $\Delta E \tau = \hbar$, i.e. $\Delta v = 1/\tau$, where τ is the lifetime. The natural linewidth is usually very low and is approximately equal to 10^{-5} nm. In practice, $\Delta \lambda$ is much higher and is caused by the Doppler effect (10^{-3} nm) and Stark splitting (10^{-1} nm).

All these facts lead to a significant broadening of the spectral width of the laser radiation. Characteristics of some light sources are given in Table 3.4.

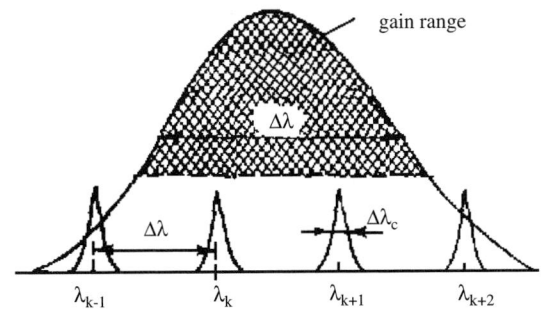

Fig. 3.6. Scheme of simultaneous amplification of longitudinal modes.

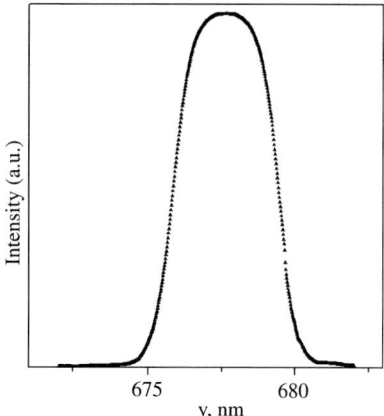

Fig. 3.7. Irradiation spectrum of semiconductor laser used in measurement setup.

Table 3.4. Parameters of light sources

Laser type	Wavelength (μm)	CW power (W)	Divergence angle (mrad)
He–Ne	0.6328	0.001–0.1	0.2–1
Ar^+	0.5145	1–10	0.5–1
	0.4880		
Kr^+	0.6471		
	0.5681	0.5	0.5–1
	0.5208		
	0.4762		
Ruby	0.6943	1	1–10
Glass with Nd	1.06	1–250	2–20
GaAs	0.84–0.9	1–10	$20 \times 400^*$

*In directions of two-coordinate axis.

Therefore, we can note that the optical quantum generators or lasers usually have a fixed radiation wavelength, although nowadays generators with tunable wavelength are found. Radiation usually takes place at some close frequencies or in some frequency range, which depends on the mode spectrum. The output power of the laser is the energy characteristic and is calculated using the integral of the power flow, which is determined by the Poynting vector in the area, limited by a light beam. In most cases laser irradiates almost a parallel beam, which is still characterized by some angular divergence. The spatial distribution of the radiation can be characterized by the distribution power density in the cross-section of the beam. These parameters cannot be determined unambiguously with high precision for all lasers; they should be measured in each case using different techniques.

As mentioned above, the laser beam often has the Gaussian distribution of intensity over the cross section; hence this shape of the light beam will be used in further description of measurements. Note that the Gauss function is Fourier-invariant. Besides, the Gaussian beam is the best with respect to divergence. We will consider the propagation of such beam in space. Let a_0 be a minimal beam size; so in the distance $z \gg z_0$ its size will be

$$a = \frac{\lambda}{a_0} z, \qquad (3.1.6)$$

where a is usually determined by the level $1/e$ of the value of the field at the axis.

It is evident from this expression that if the light beam "stretches" its Fourier transforms "constricts" and vice versa. The value of λz is the measure of the spatial extension of Fourier transforms. When propagating in space the beam retains its shape and remains a Gaussian beam. The dependence of the amplitude of the wave propagating in the direction of the z-axis on off-axis distance ρ is described by the expression

$$A(\rho, z) = \frac{a_0}{a} e^{-ik\rho/R_g} e^{-(\rho/R_g)z}, \qquad (3.1.7)$$

where R_g is the radius of curvature of the Gaussian-beam phase front.

Cross sizes of the beam are concerned with the distance along the propagation direction by the expression

$$\left(\frac{a}{a_0}\right)^2 = 1 + \left(\frac{z}{z_0}\right)^2. \qquad (3.1.8)$$

The divergence of the Gaussian beam is equal to

$$a = \frac{\lambda}{\pi a_0}. \qquad (3.1.9)$$

While focusing such beam by a spherical lens with focal distance f and by taking into account the condition $f \gg z_0$ the beam width in the focal plane can be calculated substituting (3.1.9) into (3.1.8) [136]:

$$a_0 \approx \frac{\lambda f}{\pi a}. \qquad (3.1.10)$$

If we define the area of focusing or a zone of beam waist as the region around the beam neck, where the spot size $a(z)$ increases $2^{1/2}$ times in comparison to the a_0 value, then we can find the length of the beam waist as [129]

$$z_R = \pi a_0^2 / \lambda.$$

This range stretches at both sides of the beam waist within the distance z_R. The value of z_R depends on the beam width and the focal distance of the lens used in the experiment. While focusing the radiation of the single mode He–Ne laser by the glass lens with $f = 60$ cm the length of beam waist is equal to ~ 1 cm.

The main characteristic of the source, which distinguishes lasers from thermal sources, is high coherence of radiation. The coherence is the characteristic of electromagnetic radiation that appears in the interference phenomenon and is caused by the correlation between the parameters characterizing the electromagnetic field. As coherence plays an important role in our measurements, we will consider it in more detail. The light intensity at some point is obtained as a result of superposition of two waves with phase difference φ and described by the expression

$$I = I_1 + I_2 + 2\sqrt{I_1 I_2} \, \mathrm{Re}(\gamma_{12} e^{i\varphi}).$$

where γ_{12} is referred to as the power of the radiation coherence. The absolute value of this quantity lies in the range 0–1; 0 and 1 correspond to the absolute absence of coherence and complete coherence of the light, respectively. Both these values are unattainable in practice, it is possible only to approach them. If the value of γ_{12} lies in the range 0–1, the light is partially coherent. If this value exceeds 0.88 the radiation is almost coherent. When the measurements are performed the aperture of light beam is to be restricted until the light becomes almost coherent. When $t = 0$, we can speak about spatial coherence, that is of the correlation between electric fields in different points of the wavefront at the given time. If we consider γ_{12} as a time function, there is the time coherence where we should analyze the field fluctuation in one spatial point. For example, the radiation of solid-state ruby lasers is coherent within the face plate of the ruby stick and is coherent within more than 85 ns, for a single mode He–Ne laser $\gamma_{12} > 0.9985$ [152]. In general, if the variation of the field from one spatial-time point to another has a deterministic behavior, the wave is coherent.

The coherent radiation can be considered as waves irradiated by a point monochromatic source. However, in this case the Gaussian beam with ω_0 cannot be nominally considered as a beam generated by the point source, and therefore by a coherent source. But at the same time this beam will have characteristics of coherent radiation, since it is monochromatic. For this reason a phase change at the wavefront will be constant in time, and as a result the terms of the time and spatial coherence are introduced. If the source would monochromatic, it radiates an interminable sinusoid. In practice, atoms irradiate a wavetrain with a length Δl and spectral width $\Delta \lambda$. The coherence length Δl can be evaluated from the expressions

$$\Delta\omega\Delta t = 1, \quad \Delta z\Delta K = 1,$$

where $\Delta K = 2\pi\Delta\lambda/\lambda^2$.

The coherence length

$$l_c = \frac{\lambda^2}{\Delta\lambda}, \tag{3.1.11}$$

and coherence time

$$\tau_c = l_c/c = \lambda^2/\Delta\lambda c.$$

If the spatial size of a distant radiation source is a and atoms irradiate light in all directions, at the focusing of such radiation by the lens in its focal plane one can observe a larger spot due to light diffraction. We can improve the situation by putting a diaphragm in this plane with diameter d, satisfying the condition

$$d << \frac{\lambda f}{a}.$$

when $a = 0.5$ cm, $f = 1$ m, and $\lambda = 0.6$ μm the size of such diaphragm is about 100 μm. But in practice the intensity distribution over cross section of the light beam differs from the Gaussian one. This may be caused by conditions of the laser generation or by the presence of dust and heterogeneity in the optical highway. All these conditions break the homogeneity of intensity distribution over cross section of the beam and lead to significant errors while performing measurements. The situation can be changed using a collimator with a point diaphragm put into focused on the first lens. This diaphragm makes spatial filtration of light waves diffracted on the heterogeneities mentioned above. The point diaphragm can be made of aluminum or copper foil with a thickness of ~ 10 μm. This diaphragm is placed in the focal plane of a microscope with a magnification $\sim 20 \times$ to $40 \times$, and served as the first lens of the collimator. To adjust the diaphragm in the focal plane of the microscope objective they should be put onto miniature drivers. One of the miniature driver should be a two-coordinate driver (a table with orthogonal directions of moving). Positioning of this system is performed in the following way: the system is illuminated by the laser radiation in the direction of the microscope objective and then one can watch the intensity distribution of light, passing through the diaphragm, on the screen situated beyond the diaphragm. At first the low-intensity light spot is visible, which is shifted to the center by a two-coordinate miniature drivers. The microscope objective is then moved in the direction of the location of the diaphragm with the help of second miniature driver. When the point diaphragm matches with the objective focal spot we observe a quick increase in the brightness of the light spot. If the distance between the objective and the diaphragm is reduced, the spot brightness decreases. The correlation between the magnification of the applied microscope objective and diaphragm diameter is shown below [156]

Magnification	2 ×	5 ×	10 ×	20 ×	30 ×	40 ×
Diameter (μm)	50	25	25	15	10	10

The use of such techniques allows one to form light beams that are distortionless and monotonous in the intensity distribution. It also increases the coherence of the used radiation, but leads to a significant power loss. Unfortunately, this is unavoidable because the recorded intensity distributions in our measurements are the result of the interference. Therefore, the coherence length is an important characteristic. For example, for the interference filter, passing through the spectral band $\Delta\lambda = 10$ nm, the coherence length is of 0.03 mm only. For the He–Ne laser $\Delta\lambda = 0.1$ nm, the coherence length is equal to several millimeters. There are situations in practice when l_c achieves tens of centimeters. For example, the Spectra Physics laser (119, USA) has a coherence length of about 100 m. The basic characteristics of the produced gas lasers are shown in Table 3.5. The coherence length for a semiconductor laser is equal to hundreds of micrometers, and for lumino-

Table 3.5. Basic characteristics of produced gas lasers

Lasers	Laser type	Operating wavelength (nm)	Power (mW)	Coherence length (m)
LGR 7649 (Siemens)	He–Ne	632.8	1.5	1.2
LG-38	He–Ne	632.8	40	0.2
LGN-303	He–Ne	632.8	2	1
LG-31	He–Cd	441.6	10	0.3
LG-106	Ar	488	1000	< 0.06
Spectra Physics 125A	He–Ne	632.8	50	0.2
Laser Science 220	Ar	441.6	10	0.06
RCA-2135	He–Cd	325	2	0.3

Table 3.6. Parameter of laser diode

Compound	Wavelength (nm)	Output power (mW)	Beam diverg. (deg)	Spectral width (nm)
AlGaInP	635–690	5–100	10 × 30	2–4
GaAlAs	755–880	5–100	10 × 30	1.5–2
InGaAs	960–990	50–100	10 × 30	1.5
InGaAsP	1270–1330	5–15	20 × 35	3

diodes it is tens of micrometers. Parameters of some laser diodes are given in Table 3.6. We should mention that some light sources, even thermal sources, irradiate very narrow spectral lines, and it is possible to obtain the radiation with a coherence length of about several millimeters while selecting these lines.

3.2. Techniques and Setup for the Measurement of Light Beam Intensity and its Spatial Distribution

While applying of the measurement techniques described in this book one should know the radius of the light beam (beam width), and also the intensity distribution of the radiation over the beam cross section. While studying the nonlinear optical properties by waveguide techniques, as well as by other techniques, one should know the absolute value of the light intensity affecting the sample. To determine the laser radiation intensity one should, first of all, measure the size of the spot where the light beam is focused by the lens. The lens has the radius r and the focal distance f. In this approach, when a primary divergence of the light beam is caused by the irradiation diffraction on the laser aperture, and the distance from the laser to the lens is so small that the beam is not strongly expanded, the expression for the spot size (where laser radiation is focused) is given by

$$a = f\theta, \qquad (3.2.1)$$

where θ is the beam divergence.

This expression is usually used as an empirical relationship for the evaluation of the width of the focused light beam. When the laser functions in the single-mode regime the intensity distribution in the focal plane will be (in one of the directions, which is normal to the light propagation)

$$I = I_0 \exp(-y^2/a_0^2),$$

where I_0 is the incident radiation intensity in the center of the focal spot and $I_0 = P_t/\pi a^2$, with $a = \lambda f/2\pi r$, and P_t the power of the light beam.

Let us evaluate the intensity of the He–Ne laser, which is usually used while performing measurements, with the output power of 10 mW and the angular divergence of about 10^{-4} rad. The laser radiation is focused by the lens with the focal distance $f = 1$ m. As discussed above, the lens focuses the beam into the spot with the size equal to 10^{-4} cm^2 and the intensity in the focal plane of the lens will be ~ 10 W/cm^2.

In some cases we will use glow lamps as the light source. If the radiation of such lamp is focused with the lens considered above and situated at the distance L from the lamp, and the intensity distribution is described by the Lambert law, then the light power in the focal plane will be $1/4I_t(r/L)^2$. One usually tends to obtain the uniform illumination of the lens. Therefore, if d is the size of the radiating area, the light intensity in the focal plane of the lens is given by [152]

$$I = \frac{I_t}{4\pi d^2}\left(\frac{2r}{f}\right)^2. \qquad (3.2.2)$$

Let us evaluate the radiation intensity of the lamp with an electrical power of 100 W and with the size of the radiating area of $0.1\,\text{cm}^2$, when it is focused by the lens with $r = 0.05$ m and $f = 0.02$ m. Using the expressions stated above, we can obtain that at the efficiency of the lamp equal to 5%, the light intensity in the focal plane will be equal to $1\,\text{W}/\text{cm}^2$. In general, to determine the incident light intensity one should know the light power and the size of the spot where the light beam is focused at the surface of the sample, under investigation.

The wide variety of photodetectors permits to solve the problems associated with the accurate determination of optical radiation parameters correctly [143]. Of course, there are some peculiarities evolved in such measurement techniques. The photodetector is saturated at high light powers, thus it is necessary to attenuate the power of the input beam. The use of the gray glass filter helps the photodetector to functioning in the linear regime. But it should be noted that at high light intensity the absorption coefficient of filters can also be changed. Hence one should calibrate them at the same power conditions as in experiment [152]. Gray glass filters should be used at the average intensity of approximately near $100\,\text{mW}/\text{mm}^2$. The filters as well as the photodetectors can be calibrated using the standard light sources or by calorimetric measurement [12]. Calorimetric measurement is a way of determining the laser pulse energy or CW radiation power. Calorimeters, used for measuring laser, consist of absorbers with small heat capacity, which is similar to a blackbody and the device for measuring temperature, which is in contact with the absorber. In practical applications, the calorimeter or the powermeter on its base is used for calibration of the photodetector that determines the light beam parameters. The scheme of such experiment is standard. The incident light beam is split into two parts, one of them is directed to the standard powermeter and the other is directed to the photodetector. Having performed a series of measurements with gray filters one can calibrate the photodetector.

Another problem, that appears at the laser beam parameter measurement is the wide range of the change of the radiation power. This creates the problem of the detection of low-intensity light in the presence of noise, i.e., there is need to take into account the noise of the detector and the amplifier. Such simple methods as the measurement of the ratio of a reference channel power and of a signal channel power or the power leveling for a radiation source; or more complicated methods such as the heterodyne reception [131] can be used for the recording of low-intensity signals. It is often required that in experimental setup the output power of source is constant during quite a long period. Therefore, let us consider the problem of power leveling in more details. The CW laser radiation intensity is not constant but it undergoes stochastic fluctuations. These fluctuations can be caused by instability of supply voltage, gas charge, mirror vibrations, temperature drift of resonator parameters, and so on. All these effects lead to the appearance of the noise that decreases the signal-to-noise ratio and results in additional mistakes when the thin-film parameters are measured. To eliminate these undesirable effects one can use laser power stabilization with the help of the feedback controlling the power of a laser pump. In this case some part of output radiation is directed to the photodetector (Figure 3.8).

The output signal from the detector is compared with the same output signal taken in a previous time, or with the reference signal. It is then amplified and fed to a laser power source, where it controls the discharge rate. The operating frequency band of such system is defined by capacities and by inductances in the power source. Hence, it is impossible to eliminate the rapid fluctuations of gas discharge with the help of such a setup. But this stabilization technique is quite suitable for applications considered below.

While measuring thin-film parameters we should know the width of the probe light beam, therefore, the error of the spot size measurement will affect the measured optical parameters of the studied thin films. Certainly, one can determine the beam width with the help of expressions (3.1.3)–(3.1.10). But since we are concerned with precision in measurements, it will be more correct to record the spatial distribution of the light-beam intensity over the radial coordinate by suitable equipments. Further

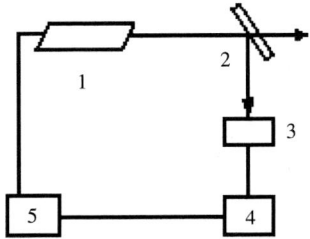

Fig. 3.8. The laser intensity stabilization scheme: laser (1), beam splitter (2), photodetector (3), comparator (4), power source (5).

approximation of the obtained dependence by some function, and also calculation of the beam width should be done. We can check the correspondence of the intensity distribution of the used light beam with the Gaussian beam. Two general schemes for measuring radiation intensity distribution can be marked out. In the first scheme a photodetector, that consists of a large number of discrete photoelements is used [153]. Certain elements of the detector are activated at the same time to record the distribution of the intensity in space. This scheme of parallel recording is convenient for recording of the field pattern in real time, but one should take into account the differences in sensitivity of the photoelements [133]. In the scheme of consecutive measurements, only one photodetector is used that is shifted to a definite section of the field pattern to be recorded. In this scheme, it is possible to obtain the intensity distribution for a quasi-stationary case. We can perform the measurements in local area by using a diaphragm with a small aperture in front of the photodetector [154]. Thus we can see that the essence of the measuring technique and that of the technique applied for testing the total power of a light beam are similar especially while using consecutive recording. But if we use this technique to measure the spatial distribution of intensity in a weak, divergent (or focused by a long-focus lens) laser beam, then we have to use high-precision mechanics for positioning the photodetector. It would be more appropriate to use photodetectors of a CCD-array type for measuring the intensity distribution.

To perform such measurements one can use a device schematically represented in Figure 3.3. The diode array (7) is used as a photodetector. In general, optoelectronic converter consists of a diode array, control unit, and a standard interface circuit that provides direct access to the computer memory. The intensity distribution in the light beam cross-section, which is recorded in the focal plane of the lens (6) using the diode array, is shown in Figure 3.9.

3.2.1. Mathematical Processing the Recorded Spatial Distribution

While performing experimental measurements one can face the negative influence of noise on the spatial distribution of the intensity of the reflected radiation. The noise caused by different sensitivities of separate cells of the matrix photodetector can be accounted by a simple subtraction of a control function. It is more difficult to fight with noises caused by the inhomogeneity of the laser beam, by the defects in the elements of the optical scheme, asperity of surfaces of the investigated structure and of the prism coupler. All this leads to the recording of a noisy signal that reduces the precision of determining thin-film parameters. The recorded intensity distribution (see Figure 3.9) is a series of experimental data I_i and y_i. Generally both the measured quantities are characterized by measurement errors. Thus, their processing requires application of mathematical optimization methods. In order to determine the coordinates of extremum of the recorded distribution with a minimal error it is advisable to use the functional approximation of the series of measured values I_i and y_i ($i = 1, \ldots, l$) by the mathematical model $I = f(y)$, where y is the vector containing m parameters. Consideration of such model leads to the system of m equations $I_i = f(y) + \zeta_i$, where ζ_i represent a misclosure. Choice of the elements of the vector y, when $l >> m$ (more often $l \geq m + 2$), is performed in order to minimize the sum $F = \sum_i \zeta_i^2$. When function $f(y)$ is linear in y, we have the classical least-squares problem [164,165]. This approach will be used further, and so we will consider its basic elements. Let us consider the quantities y_i as independent variables, the values of which have been measured with a small error, and I_i corresponding to y_i deviates from the correct value $I(y_i)$ on $I_i - I(y_i)$ (Figure 3.10).

Then the best straight line $I = ay + b$ is the line, which has the minimum of the sum of deviation squared $\sum_{i=1}^{N}(I_i - ay_i - b)^2$. As the condition for the minimum of F is that the value of the first derivatives is equal to zero, so

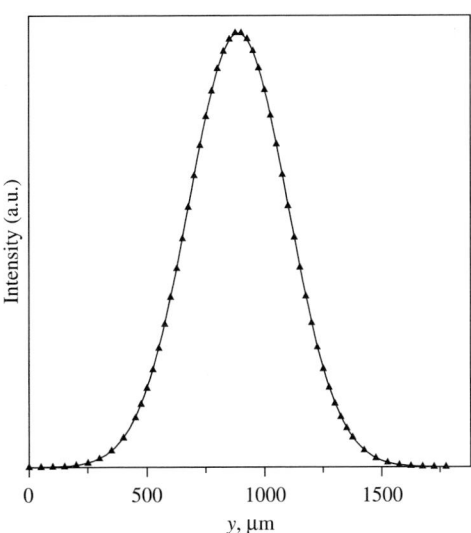

Fig. 3.9. Intensity distribution in the cross-section of the light beam along one of transverse coordinates.

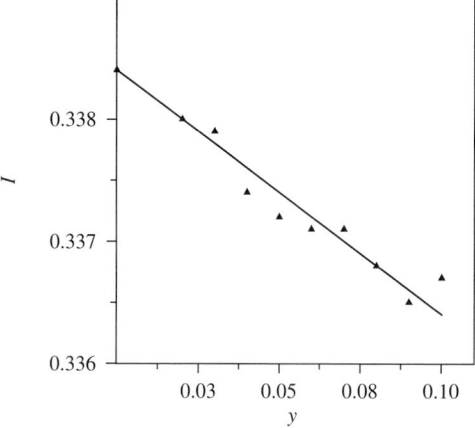

Fig. 3.10. Experimental values and their approximation by least-squares method.

$$\sum_{i=1}^{N}(y_i I_i - a y_i^2 - b y_i) = 0,$$

$$\sum_{i=1}^{N}(I_i - a y_i - b) = 0.$$

From the last set of equations one can derive unknown variables a and b as

$$a = \left(\sum_i y_i \sum_i I_i - N \sum_i y_i I_i\right) \bigg/ \left(\left(\sum_i y_i\right)^2 - N \sum_i y_i^2\right),$$

$$b = \left(\sum_i y_i \sum_i y I_i - \sum_i I_i \bigg/ \sum_i y_i^2\right) \bigg/ \left(\left(\sum_i y_i\right)^2 - N \sum_i y_i^2\right).$$

Taking into account the arithmetic mean of y_i and I_i,

$$\bar{y} = (1/N)\sum_i y_i, \quad \bar{I} = (1/N)\sum_i I_i,$$

the equation of line will be $\bar{I} = a\bar{y} + b$. The obtained line will pass through the mean values of y_i and I_i (see Figure 3.10). The dispersion $\sigma_n^2 = F_{\min}/(N-2)$ characterizes a spread in I_i near the mean value.

When one has to find a nonlinear function $f(y)$ analytically, there is a problem of nonlinear optimization. These are problems, which usually appear at the processing of the intensity distribution of the reflected light beam. If the function has one extremum on the specified interval, one needs

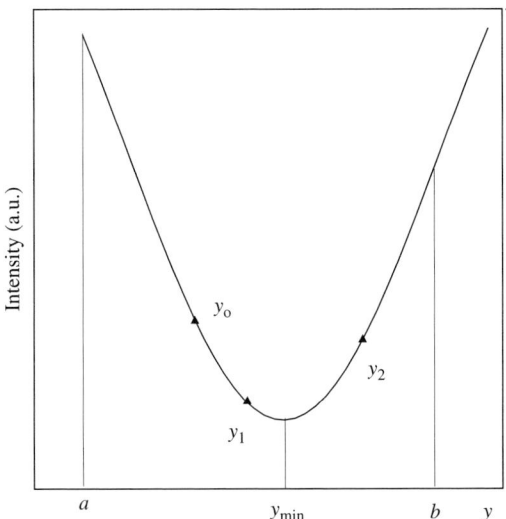

Fig. 3.11. Intensity distribution near the minimum.

to determine the values of the function at three points within the interval in order to localize the coordinate of the extremum. In order to determine the minimum of the function two general classes of algorithms are used. The first is based on the calculation of function values optimally chosen from sampling points but the behavior of function is not accounted. Furthermore, information about the function behavior can be used for the evaluation of the function extremum and for the choice of the next approximation. Interpolation formulas are used for the approximation of the function and the values of minimum or maximum are obtained by differentiation of these expressions [166]. We can cite another algorithm used to find the minimum of function. If the value of function $f(y)$ is given at point y_0, the values $f(y_0 + h)$ and $f(y_0 + 2h)$ are calculated (Figure 3.11).

If these three points do not allow to localize the minimum, so the calculations are performed at points $y_0 + 4h$, $y_0 + 8h$, until the minimum is localized. The interval is then halved. If this search does not give a result, the process is performed in the opposite direction. Knowing the coordinates of three points y_0, $y_1 = y_0 + h$, $y_2 = y_0 + 2h$ (see Figure 3.11), one can formulate an interpolated quadratic polynomial

$$f(y) = a + by + cy^2.$$

Since the parabola has the minimum at point $y_{\min} = -b/2c$, then

$$c = f_0/(y_1 - y_0)(y_2 - y_0) - f_1/(y_1 - y_0)(y_2 - y_1) + f_2/(y_2 - y_1)(y_2 - y_0)$$

$$a = f_0 - by_0 - cy_0^2,$$

$$b = f_0(y_1 + y_2)/(y_1 - y_0)(y_2 - y_0) + f_1(y_0 + y_2)/(y_1 - y_0)(y_2 - y_1)$$
$$- (y_0 + y_1)/(y_2 - y_1)(y_2 - y_0).$$

The second method is the method of gradient descent or steepest descent, proposed by Koshi [189]. It is one of the nonlinear optimization methods based on the use of the derivatives of the analyzed function. It allows one to find solutions of the nonlinear equation system using minimum of required calculations. The essence of the method can be explained in the following way. Let the function $f(X)$ at some point X_0 have the value $f(X_0) = b$, where X, in general case, is the complex quantity. The series of points, where function $f(X)$ is equal to b, forms a line or a surface of equal level. To determine the extreme values of $f(X)$, we have to move from this surface in the direction of the greatest change in the values of this function. One should use the direction opposite to the gradient, or the direction of steepest descent, as the search direction. If X_0 is the initial approximation, then the sequence of iterations is described by the expressions

$$\Delta X_i = -\alpha_i \nabla f(X_i), \quad X_{i+1} = X_i + \Delta X_i,$$

where α_i is the scalar factor, minimizing the value of $f[X_i - \alpha_i \nabla f(X_i)]$.

While using the method of gradient descent in practice one should pay attention to the fact that the result depends on the variations of the independent variables. This is owing to the fact that the lines of the equal levels are mostly oblong and are not circles (the method of gradient descent converges after one step for circles) and the result of the search depends on initial approximation. If the latter is specified with high error, the application of this method would not give a positive result [168]. In this way, using one of the given methods of optimization accurate to round-off errors occurring in calculations, coordinates of the function extremum are determined.

3.2.2. Determination of the Light-Beam Parameters

The approximation of the measured distribution by Gaussian function and its further processing allows one to determine the beamwidth a_w in the plane y', i.e., where the prism coupler will be placed during the measurement of thin-film parameters. We will use the method of Fourier analysis for this. Such approach is stated in more details in Chapter 4.

Let $f(y') = \exp(-y'^2/a_w^2)$ be the spatial distribution of an incident light beam intensity (see the geometry of problem in Figure 3.2), then the Fourier spectrum of the recorded ligsht beam can be written as [160]

$$\Psi = \frac{1}{2\pi} \int_{-\infty}^{\infty} e^{ik_y y' - (y'/a_w)^2} \, dy';$$

It is easy to show that

$$\Psi = \frac{1}{\pi} \int_{0}^{\infty} \cos(k_y y') e^{-(y'/a_w)^2} \, dy' = \Psi_0 e^{-k_y^2 (a_w/2)^2}, \tag{3.2.3}$$

where Ψ_0 is some constant that is not of interest at the moment, $k_y = k_0 \sin(y/f)$, $k_0 = 2\pi/\lambda$, y', y are coordinates, and f is the focal distance of lens (6).

In the approach of paraxial optics $k_y = k_0 y/f$. If we set such intensity level that we can determine beam radius as $I_a = I_0/e$, then from (3.2.3) the following expression can be obtained:

$$a_w = \frac{2f}{k_0 y_a}, \tag{3.2.4}$$

where y_a is the recorded beamwidth (Figure 3.9) at the level $I_0 e^{-1}$ in the focal plane of lens 6.

The value of y_a can be easily found using the approximation of the experimental data $\Psi(y_i)$ by the Gauss distribution, because it is easy to form the Gauss distribution for a beam with a typical size y_a. Further, using the least-squares method during the condition of the minimization of the misclosure $\partial \xi / \partial A = 0$, one can find A and calculate y_a.

Here $\sum_{i=1}^{m} [\Psi(y_i) - A \exp(y_i/y_a)]^2 = \xi$, and m is the number of experimental points in the recorded distribution. Analysis of the data, depicted in Figure 3.9, shows a satisfactory correlation of the obtained experimental intensity distribution (it is depicted by discrete points in the picture) and Gauss distribution (continuous curve). The beamwidth is equal to 147.5 μm for the given distribution. The value of σ^2 is used as an experimental evaluation of dispersion at the description of the measurement results

$$\sigma^2 = \frac{\sum (a_i - a_w)^2}{(N-1)}, \tag{3.2.5}$$

where $i = 1, \ldots, N$, $a_w = \sum a_i/N$ is the averaged value of the recorded data a_i for the series of N measurements.

It is obvious that σ^2 is an asymptotically unbiased estimator of dispersion. Then the interval $(a_w - (\sigma/\sqrt{m}) t_{\alpha,m-1}, a_w + (\sigma/\sqrt{m}) t_{\alpha,m-1})$ is the confidence estimate of the mean value of measurements (with probability P_α), where m is the number of series of measurements and $t_{\alpha,m-1}$ is the Student's coefficient.

Then, the maximal value of the absolute measurement error will be $\delta = \sigma t_{\alpha,m-1}/\sqrt{m}$, and the relative error will be defined as $\delta_R = \delta/a_w$. For the given example of determination of a_w by the series of 50 measurements $\delta = 2\sigma = 0.04$ μm.

Therefore, having studied the characteristics of light-beam by the technique stated above, we can start measuring the parameters of thin-film structures.

3.3. Spatial Distribution of the Light Beam Intensity Reflected from the Prism Coupler and Measurements of Thin-Film Parameters

When the guided modes are excited using the traditional prism-coupling technique, as already noted, one can observe a series of dark *m*-lines in the reflected light. If a separate line is examined carefully, one can reveal that the *m*-line has its own "fine" structure (see Figure 3.1). This structure is well marked for thin-film waveguides having small optical losses. This intensity distribution can be recorded at the appropriate installation of the setup parameters. The setup, which is used for experimental observation and recording of the spatial distribution of the reflected light beam intensity in the case of the excitation of guided modes in thin-film waveguides, is shown in Figure 3.12.

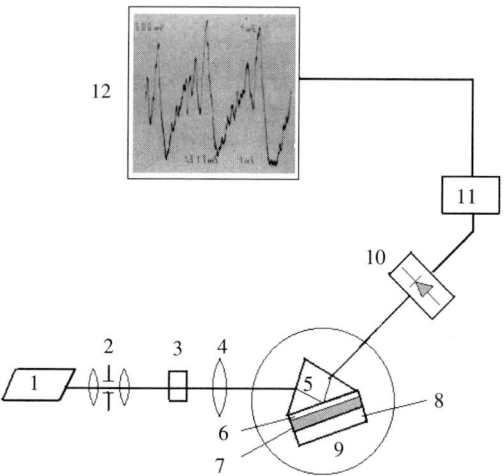

Fig. 3.12. Experimental setup used for studying of the spatial distribution of reflected light beam intensity in case of excitation of the guided mode using the prism coupler: light source (1), collimator (2), polarizer (3), lens (4), prism coupler (5), gap (6), waveguide (7) on substrate (8), rotary table (9), photodetectors (10), intensity meter (11), display (12).

The setup is constructed on the basis of a goniometer. An electronic circuit for the measurement of the intensity distribution parameters consists of optoelectronic converter (11) and CCD array (10). The CCD array is connected with a digital oscillograph (12). A photodetector (10) is positioned at a certain distance z from the output face of the prism (5). In the example considered, $z = 0.74$ m. The rotation axis of the CCD array is positioned in a line with the rotation axis of table (9) of the goniometer. The He–Ne laser (1) with a radiation wavelength of 632.8 nm is used as the light source. After passing through the collimator (2) and the polarizer (3), the laser radiation is focused by the lens (4) with diameter of 3 cm. The light beam has the beam $a_w = 30$ μm. The beamwidth a_w is determined by the intensity level $I = I_0 e^{-1}$, where I_0 is the intensity at the beam axis when the intensity distribution is recorded in the focal plane of the lens (4). The typical structure of the one-dimensional distribution of the radiation intensity over the cross-section of the reflected beam in the case of the excitation of the single guided mode (in direction normal to the observable m-line, see

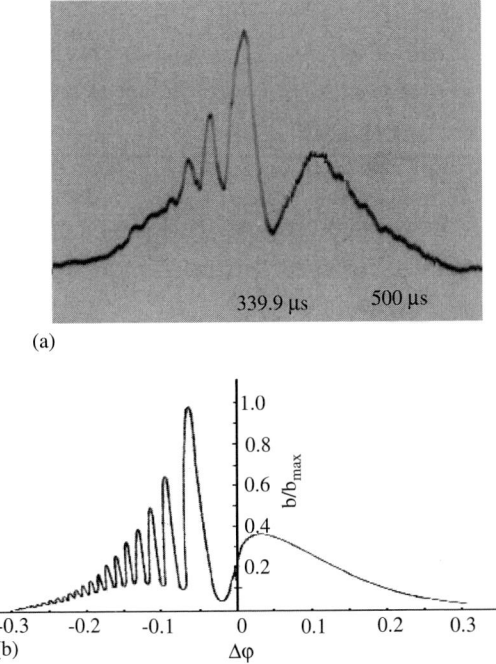

Fig. 3.13. Spatial distribution of the intensity of the light beam reflected from the prism coupler in case of the guided mode excitation: recorded in experiments (a) and calculated (b) on the basis of the proposed model [126].

Figure 3.1, section A-A), is given in Figure 3.13a. The results of the study of thin-film waveguides with different optical properties reveal the fact that every thin-film structure has special parameters of such intensity distribution. These parameters are typical for certain films [126]. A similar intensity distribution, and the basis and interpretation of the observed effects are given by Manneret et al. [127]. However, they associate the non-monotone character of the spatial distribution with the thickness of the air gap (a layer between the prism coupler and waveguide) and they use reflected radiation spatial distribution parameters to determine the air gap thickness. We can agree with the authors of Ref. [127] that the thickness of the air gap really affects the intensity distribution of the reflected light beam. But at the same time this distribution is specific for every guiding structure. Therefore, there is the need to consider the parameters of that distribution and to associate them with the properties of the thin film. This correlation between thin-film properties and intensity distribution of parameters have been discovered. Furthermore, the information about the film material parameters (the refractive index, the absorption coefficient, and the film thickness) is "encoded" in the angular width of the minima of the recorded intensity distribution and in its contrast. This can be explained in the following way. Let us suppose the light beam is incident on the measuring prism (see Figure 3.12) and excites the guided mode in the waveguiding structure.

It is obvious that the complex propagation constant of this mode depends on the absorption coefficient, the refractive index and the film thickness. Being partially reflected from the prism base, the incident laser beam is coupled by the prism into a planar waveguide. Under total internal reflection at the prism base, the strong coupling of light into the waveguide can occur via resonant frustrated total reflection, i.e., via evanescent waves in the air gap between the prism and the waveguide. Being propagated along the thin-film structure the light beam is reradiated into the prism and interferes with the reflected beam. Thus, the output light beam contains information about the investigated sample. While determining the film's parameters the results of processing of the spatial intensity distribution of this light beam are used. The intensity distribution, which was obtained on the basis of a theoretical model proposed in Ref. [126], is depicted in Figure 3.13b and it satisfactorily fits the experimental data. The position of the m-line corresponds to the position of the first minimum situated to the right of the largest intensity maximum. The correlation of the calculated and experimental results makes it possible to use the conclusions of this study for determining the parameters of waveguided films. While processing the recorded distribution of the intensity of the reflected light beam it is necessary to measure angles φ corresponding to the intensity minimum I_{min}, to one of the maxima I_{max}. Then, it is possible to construct dependencies of A_{min} and

V on the parameter ΔA. These dependencies associate the parameters of thin-film structures and record the intensity distribution. Such typical dependencies are given in Figure 3.14. ΔA is related the experimentally determined parameters φ_a, φ_{max}, φ_{min}:

$$\Delta A = \sqrt{\frac{k_0 z}{2 n_g \cos^3 \varphi_a}} \cos \varphi_a (\varphi_{max} - \varphi_{min}), \quad (3.2.6)$$

where

$$A_{min} = A(\varphi_{min}),$$

$$A_{max} = A(\varphi_{max}),$$

$$A = Q_1 \left[\sqrt{n_p^2 - (h' k_0^{-1})^2} \sin \beta - h' k_0^{-1} \cos \beta + n_g \sin \varphi_a - n_g \cos \varphi_a \Delta \varphi - \frac{L+l}{z} \right.$$
$$\left. \times \frac{n_g \cos^3 \varphi_a \sin \varphi_a}{Q_3} \right],$$

$$V = Q_1 Q_2 h'', \qquad \varphi = -\varphi_a + \Delta \varphi,$$

$$Q_1 = \sqrt{\frac{k_0 z}{2 n_g \cos^3 \varphi_a}},$$

$$Q_2 = \frac{2 n_p \sin \theta_p \sin \varphi_a \cos^2 \varphi_a}{z(n_p^2 - n_g^2 \sin^2 \varphi_a)},$$

$$Q_3 = \sqrt{n_p^2 - (n_g \sin \varphi_a)^2}.$$

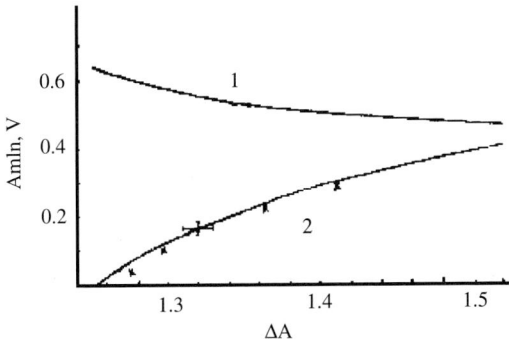

Fig. 3.14. Dependencies of A_{min} (1) and V (2) on the ΔA parameter obtained at the maximum of contrast.

In these expressions φ_a is the angle between the light beam axis and the normal to the entrance face of the prism coupler, n_p, n_s and n_g are the refractive indices of the prism, substrate and air gap, respectively, θ_p is the angle at the basis of the prism coupler, l and L are the distance from the entrance and target faces of the prism to the light-beam center on the basis of the prism coupler, respectively, z is the distance from the target side of the prism to the photodetector (zone of supervision).

The values of angles φ_{max} and φ_{min} are counted off from the normal to the target side of the prism coupler, and the parameter V depends on the contrast of the recorded picture $K = (b_{max} - b_{min})/(b_{max} + b_{min})$.

The behavior of dependencies given in Figure 3.14 is determined by the parameters of the substrate, prism coupler, and the surroundings. Here it is worth noting that the measurements should be carried out at the greatest possible contrast achieved during the observation of the spatial distribution of the reflected radiation intensity for the excited mode; otherwise optical losses will be determined with a significant error.

By using the above stated dependencies on the measured values of φ, $\Delta\varphi$ and ΔA calculated via (3.2.6) we can find A_{min} and V. Then, the real and the imaginary parts of the mode propagation constants can be determined from the following expressions

$$h' = k_0 n_p \sin\left[\left(\theta_p - \arcsin\frac{n_g \sin\varphi_a}{n_p}\right) + \frac{A_{min}}{Q_1} + Q_2\right], \qquad (3.2.7)$$

$$h'' = \frac{V}{Q_1}\sqrt{\frac{n_p^2 - n_s^2}{n_p^2 - n_g^2\sin^2\varphi_a}}, \qquad (3.2.8a)$$

Optical losses of a guided mode are

$$p = 2 \times 10^5 \, k_0 h'' \lg e \quad (dB/cm). \qquad (3.2.8b)$$

Expression (3.2.7) differs from (2.3.2) used traditionally by the last two terms. The value of these terms can be significantly reduced due to the increase in the distance between the target face of the prism and the photodetector (in this case such moving off is accompanied by the decrease in the coupling between the prism and the waveguide). The accuracy of h' determination is also increased and the error of determination of the real part of the mode propagation constant will be indefinitely small within the limit $z \to \infty$. This reconstruction of propagation constants will be correct in practice if, for example, losses in waveguide are not less than 2.5 dB/cm at $z = 1$ m and 0.8 dB/cm at $z = 10$ m.

Let us illustrate the example of the application of this approach on polymeric thin-film waveguides obtained on the substrate made from the optical glass TK14, by using the extraction from polyvinyl-teraftalate solution in "dichlorethane-chlorbenzene" mixture (1:1). This waveguide had the step-like profile of the refractive index [126]. The prism made from the optical glass TF5 with $\vartheta_p = 60.128°$ was applied for the tunnel excitation of guided modes. The angle φ was equal to 22.1°, $L = 0.56$ cm and $n_a = 1.00033$. Angles φ_{min}, φ_{max} were measured and they corresponded to the contrast maximum of the intensity distribution that was obtained by the air gap thickness variations. Measurements were performed at 0.74 m $< z <$ 9.20 m. The values $z > 0.74$ m were obtained with the help of two mirrors. For this thin-film waveguide the dependence depicted as curve 2 in Figure 3.14 was calculated for $n_c = 1.00033$ and $n_p = 1.93601$. The angle φ and distances $(L + l)$ for the different guided modes are depicted in Table 3.7. To avoid the additional errors caused by a deformation of the film due to clamps of the prism to the waveguide, it is necessary to perform the measurements with a constant thickness of the air gap.

The propagation constant values for TE modes h' obtained from expression (3.2.7) taking into account correction terms and h'^* obtained via the traditional methods [24,77] are also depicted in Table 3.7. Here, it should be noted that an acceptable contrast of displayed picture for all modes at one time was not achieved. The contrast for the main mode ($m = 0$) was less then the one required for the effective measurement of φ_{min} and φ_{max}. It is obvious that the h' values systematically exceed h'^*. The successive substitution of the measured values h' for any two guided modes into the dispersion equations for a planar waveguide (see expressions (2.2.1)–(2.2.2) allows one to determine the refractive index n and the thickness d of the waveguide. While choosing all possible pairs of h' and h'^* three values of n, d and four values of n^*, d^* can be obtained. Their averaging gives values $\bar{n}^* = 1.62207$, $\bar{d}^* = 7.10$ μm, and $\bar{n} = 1.62201$, $\bar{d} = 7.47$ μm. It is obvious that the difference between h'^* and h' values basically affects the determined value of the thickness of films.

Thus, experimental results reveal the possibility of the determination of the film's parameter using the results of the processing of the spatial intensity distribution of the reflected light beam. The experimental verification of

Table 3.7. Mode parameters of the polymeric waveguide

Modes	φ_i (deg)	$(L+l)$ (mm)	$(h'/k_0)^*$	h'/k_0
1	−6.40	13.028	1.62046	1.62046
2	−6.70	13.031	1.61733	1.61770
3	−7.03	13.033	1.61414	1.61455
4	−7.38	13.036	1.61075	1.61122

the real part of the mode propagation constant determined with the help of the stated approach is quite problematic as it requires the determination of h' by the independent method that should provide the same accuracy as the prism-coupling technique. Moreover, there are no such reliable methods to date. But the experimental determination of the waveguide losses is easier because the absorption coefficient can be measured without the prism coupler application. These measurements can be fulfilled by the photometry of the mode propagation length [93]. During the investigation of thin-film waveguides obtained by RF sputtering of quartz glass on substrates made from optical glass KV in the argon and oxygen (20%) mixture at a total pressure of 0.1 Pa, the guided modes have attenuation from 8 up to 35 dB/cm, depending on the fabrication conditions. The obtained experimental data are depicted in Figure 3.14 by dots, where V values are determined for p values measured by the photometry of the mode propagation length. The results depicted in the figures show that the calculated dependence $V(\Delta A)$ (curve 2) is in good agreement with the experimental data.

On the basis of the given results we can conclude that the spatial intensity distribution of the light beam, which was reflected from the prism coupler, contains information about the parameters of the studied thin film. Unfortunately, this approach is of little use as a measurement method, because the dependencies (see Figure 3.14) are not universal. Furthermore, these dependencies should be recalculated for one thin-film structure each time while changing the prism coupler. But the stated approach of studying the spatial intensity distribution of the reflected radiation when the guided mode is in an excitation state serves as a starting point for the development of a new generation of the waveguide methods and devices for the measurement thin-film parameters.

Chapter 4
Spatial Fourier Spectroscopy of Guided Modes: Measuring Thin-Film Parameters

4.1. Fourier Transform Applications for Studying the Spatial Distribution of the Reflected Light Beam Intensity... 77
4.2. Studying the Properties of Waveguiding Films........................... 83
    4.2.1. Features of Recording the Intensity Distribution and Measuring the Complex Propagation Constants of Modes 83
    4.2.2. Determination of Thin-Film Parameters 88

The integrated-optics methods of measuring thin-film and surface layer parameters are based on the determination of the propagation constants of guided modes, which are excited in the investigated structure. The method well known as the prism-coupling technique, which is also referred to as the m-lines technique, is commonly used to determine the optical parameters of thin films [34]. As already mentioned, the propagation constants can be found while recording the position of the resonant minimum in the angular dependence of the reflection coefficient of the light beam with the help of the prism-coupling technique. This chapter summarizes the results of the development of the waveguide methods of investigating thin films. The considered approach is also based on the recording of the spatial distribution of light beam intensity reflected from a prism coupler. While studying the guiding properties of the film the information about the angular position of the m-line (the resonant angle of guided mode excitation) only is considered. This makes possible to measure the refractive index and thickness of the film. In the case of application of the approach considered in this Chapter it uses information about the whole array of experimental dots of the intensity distribution. This makes it possible to determine the real and imaginary parts of the mode propagation constants and allows one to determine not only the refractive index and thickness, but also the absorption coefficient of thin films.

4.1. Fourier Transform Applications for Studying the Spatial Distribution of the Reflected Light Beam Intensity

While measuring the spatial distribution of the reflected light intensity in the experiments (described in Section 3.3, the changes in the intensity distribution depicted in Figure 3.15) and its transformation into a symmetric dependence (Figure 4.1) were recorded with the increase in the distance between the photodetector and the prism coupler (increase in z). This can, however, be observed if the light beam propagates through a lens, whose back focal plane coincides with the location of the photodetector.

On the other hand, it is well known that the electromagnetic field $\psi(u, v)$ in the back focal plane of the lens L_1 (Figure 4.2) is the two-dimensional Fourier transformation of the field $f(x, y)$ in the front focal plane of the lens:

$$\psi(u, v) = \frac{1}{i\lambda f} \exp\left[i\pi\lambda f\left(1 - \frac{l}{f}\right)(u^2 + v^2)\right]$$
$$\times \int_{-\infty}^{\infty}\int f(x, y) P e^{-i2\pi(ux+vy)}\, dx\, dy, \qquad (4.1.1)$$

where f is the lens focus, l the distance between the lens and the object, λ the

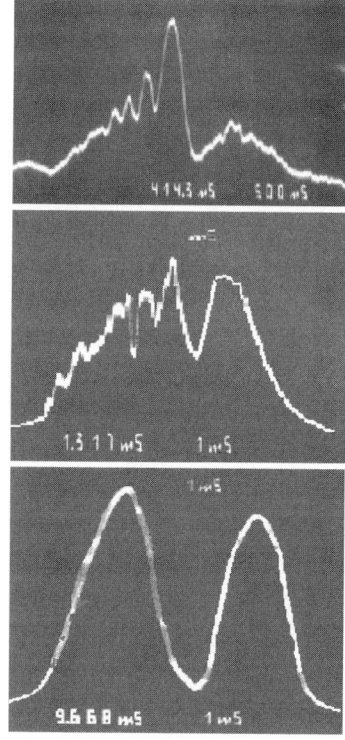

Figure 4.1. Changes in the spatial distribution of the reflected light beam intensity with the increase in the distance between the photodetector and the prism coupler: $z = 0.74, 3.0$ and 9 m, respectively.

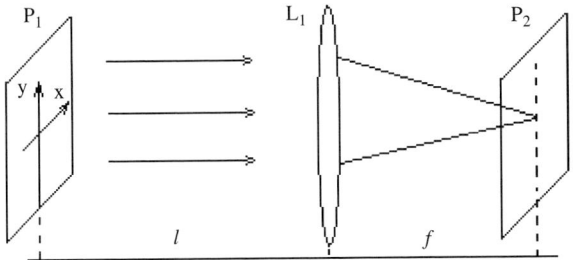

Figure 4.2. Setup of the optical system performing the Fourier transforms.

light wavelength, L the lens diameter and P is equal to 1 at $(x+y)^{1/2} < L/2$ or $P = 0$ in other cases. The coordinates x and y are related to spatial frequencies u,v in the transform plane P_2 by the expressions

$$x = fu\lambda; \quad y = fv\lambda. \qquad (4.1.2)$$

From expression (4.1.1) it follows that when the sample is placed in the front focal plane ($l = f$), the phase multiplier before the integral in (4.1.1) is equal to unity. If the influence of the pupil is neglected, the front and back focal planes of the lens are related to the Fourier transforms. But when the sample is located closely to the lens ($l = 0$), the influence of the size of the pupil practically has no effect when $D < L$, where D is the maximal size of the sample. Thereby, the Fourier spectrum is recorded in plane P_2 taking into account the approach of the zero phase multiplier. For other cases the accuracy of the transform is determined by the values of spatial frequencies u and v. Thus, when $D < L$ the lens will work as the low-pass filter. The attenuation of high-frequency components in the Fourier spectrum can be reduced by decreasing the distance l. Although the phase multiplier in (4.1.1) is not eliminated at small l and the phase shifts appear between the separate sections of the wavefront of the propagated light, it does not affect the measurement of the spectrum of intensity, which is proportional to $|\psi(u,v)|^2$. So, in practice, one prefers to place the sample as close to the lens as possible [128]. In the experiments the lens performs the Fourier transform of the light beam and its intensity distribution over spatial frequencies, i.e. the angular Fourier spectrum of the reflected light beam, is recorded [155]. *Further, in this approach we shall not discriminate between the concepts of the Fourier spectrum of the light beam and the spectrum of light intensity.*

The representation of functions by the trigonometric series suggested by Fourier at the beginning of the 19th century was considered for a long time as a mathematical technique far from reality. Only in the early 1920s the wide application of the spectral ideas began in such application areas of sciences as technical mechanics, acoustics and radiophysics. Although almost all applied optics were devoted to the processing of spatial signals, the theory of Fourier optics was adopted from radiophysics. From there also "came" into optics unusual terms such as, "spatial frequency" or "transfer characteristic". A wide application of Fourier analysis in optics began in the 1960s after the invention of laser. Since the Fourier transform has been used in all applications considered in this book, we will briefly discuss its basic characteristics. The rigorous theory requires significant mathematical calculations, but the basic representations are rather simple. When the complex signal passes through a system it is convenient to represent the signal as the sum of the elementary signals. After studying the propagation of each elementary signal, one can sum all these output signals and determine the reaction of the system on the complex signal. The Fourier analysis technique is based on such an approach. In practice, a signal is always restricted in space and time. If, for example, the intensity distribution of a light beam is a function recorded in the finite interval of spatial frequencies, then this interval should be significantly larger than some typical size of the light beam. Let us consider a one-dimensional case. For example, if $f(x)$ is the input function, its Fourier transform (when the phase factor is neglected) is given by [155]

$$\psi(u) = \frac{1}{2\pi} \int_{-\infty}^{\infty} f(x) \exp(-i2\pi xu)\, dx, \qquad (4.1.3a)$$

where u is the spatial frequency and related to coordinate x by expression (4.1.2).

The inverse Fourier transform can be written as

$$f(x) = \int_{-\infty}^{\infty} \Psi(u) \exp(i2\pi xu)\, du, \qquad (4.1.3b)$$

A closer look of these expressions show that each point of the Fourier transform plane contains contributions from all points of the front focal plane, and vice versa. If we consider a point light source and $f(x) = \delta(x)$, then $\psi(u) = 1/2\pi$, i.e. the luminous point creates the homogeneous response in the transform plane; in other words the luminous point is transformed to the plane wave in the focal plane P_2. The converse also holds: a plane wave is transformed to the point in the Fourier transform plane, i.e. $\psi(u) = \delta(u)$. Thus, the wider spatial spectrum of radiation corresponds to the smaller spatial size of the light source.

We will examine the Fourier transforms performed by a lens in more detail (Figure 4.2). Let the light beam be incident on the lens. Then

$$f(x, y) = A(x, y) \exp[-i\varphi(x, y)], \qquad (4.1.4)$$

where $A(x,y)$ and $\varphi(x,y)$ are the amplitude and the phase of the incident wave, respectively.

Analogous to expression (4.1.3a) the field in the plane P_2 for a two-dimensional case is given by [157]

$$\psi(u,v) = \frac{1}{(2\pi)^2} \int_{-\infty}^{\infty}\int f(x,y) \exp[-i2\pi(xu+yv)] \, dx \, dy, \quad (4.1.5)$$

where u and v are spatial frequencies in the plane P_2, $x = u\lambda f_1$, $y = v\lambda f_1$, and f_1 is the focus length of lens L_1.

If the second lens L_2 is placed behind the transform plane P_2 (Figure 4.3), the distribution of the light field in the target plane P_3 will be double the Fourier transform of the initial function. Thus, the magnification of such optical system is equal to the ratio of the focal lengths of lenses L_2 and L_1.

The given expressions can be explained from the example of the plane wave propagating in the direction of the angle γ to the axis z, which is normal to the recording plane [158]:

$$f(\bar{r}, t) = e^{(i\omega t)} e^{(-i\bar{m}\bar{r})},$$

where $\bar{m}\bar{r} = (m \cos \alpha)x + (m \cos \beta)y + (m \cos \gamma)z$, $\bar{m}$ is the unit vector and α, β, γ are angles between the vector $\bar{m}$ and coordinate axes.

Then in the plane $z = 0$,

$$f(x,y,0,t) = e^{i\omega t} e^{-i(m_x x + m_y y)}.$$

For the point source the exponent in expression (4.1.5) is written for the point with coordinates (a,b) as

$$e^{-i2\pi(au+bv)} = \exp\left(-i\frac{2\pi}{\lambda}\left(\frac{ax}{f_1} + \frac{by}{f_1}\right)\right)$$

From the analysis of the last two expressions it follows that the exponent in expression (4.1.5) can be considered as the plane wave with m_x and m_y, i.e. the Fourier integral represents the field of a series of plane waves propagating in different directions. The function $\Psi(m_x, m_y)$, being the Fourier

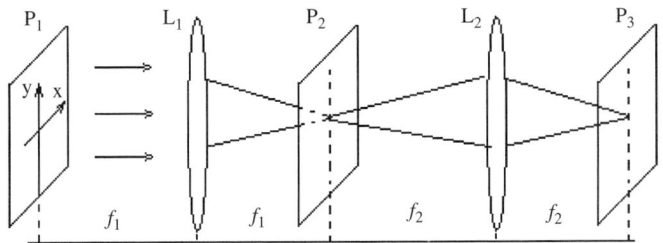

Figure 4.3. Setup of optical system performing double Fourier transform.

image of $f(x, y)$, is referred to as the angular Fourier spectrum of the field $f(x, y)$. Further, we will use the term "Fourier spectrum" to denote light beam intensity spectrum in the wavevector space (the spatial frequency decomposition of the light field).

When the guided mode is excited, the application of the Fourier transform method in investigating the spatial distribution of the reflected light beam can be illustrated as follows: If in the plane P_2 (see Figure 4.3) the opaque screen is placed on the optical axis, the low-frequency components of the input signal will be discarded and the radiation with higher spatial frequencies will pass. The distribution of the passed field obtained in plane P_3 will be "filtered". Now, if the thin-film waveguide excited by the prism coupler is used as the "filter", similar results are obtained. The light beam that partly reflects from the prism base excites the guided mode in the film due to the coupling the radiation. In the process of propagation along the thin film the light is radiated into the prism and interferes with the light beam reflected from the prism base. We can observe of the intensity distribution in the focal plane of the lens, through which the reflected light beam passes. The intensity spectrum of an incident light beam is depicted in Figure 4.4 (curve 1). But as some part of spatial frequencies is coupled into the waveguide in the case of the guided mode excitation and is partly absorbed in the process of the light propagation along the waveguide, the resulting spectrum of the light beam looks like the curve depicted in Figure 4.4 (curve 2). The absorption of the light in the guiding films appears as pronounced minimum in the spatial Fourier spectrum of the light beam

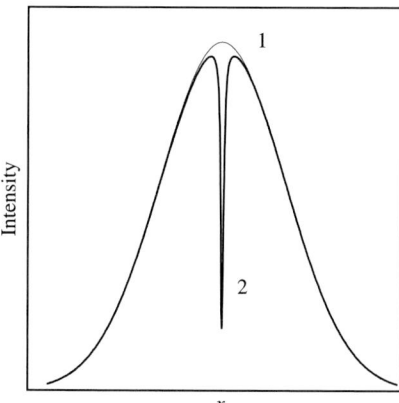

Figure 4.4. Fourier spectrum of the light beam reflected from the prism coupler without the waveguide (1) and in the case of excitation of the guided mode in a thin-film structure (2).

reflected from the prism coupler at the excitation of guided mode in thin-film structures.

However, it should be noted that the last statement is true if the radiation does not go beyond the prism coupler, i.e. the mode track is smaller than the size of the prism's base.

4.2. Studying the Properties of Waveguiding Films

This section deals with the further development of the waveguide methods of investigating thin film. As shown in the previous chapter, the symmetric picture of light intensity distribution is recorded when the light beam reflected by the prism coupler propagates through the lens, whose focus plane coincides with the location of the photodetector. This distribution recorded in the focal plane of the lens is the angular Fourier spectrum of the light beam or the intensity spectrum of the light beam [158]. As it follows from the experiments, the parameters of such spectrum depend on the properties of the thin-film structure, where the guided mode is excited. It has been shown [160–162] that the appropriate processing of Fourier spectra of the light beam recorded when the guided modes are excited allows one to determine the parameters of thin film.

4.2.1. Features of Recording the Intensity Distribution and Measuring the Complex Propagation Constants of Modes

The spatial distribution of intensity, which is observed in the cross-section of the reflected light beam when the guided mode is excited by the prism-coupling technique [159], can be recorded by the experimental setup [160] shown in Figure 4.5. If we take into account the conclusions of Section 3.1, then it would be appropriate to choose a single-mode gas laser as the source of radiation. The He–Ne laser with a radiation wavelength of 633 nm is used as the light source in the considered setup.

The one-dimensional spatial distribution of the reflected light beam intensity $I(y)$ containing information about the film parameters is measured by the CCD-array photodetector (12), which is connected to the digital image processing system (18). The rotation axis of CCD array (12) is connected to the rotation axis of the spectroscopic goniometer table (6). After a digital processing the signal comes to the random-access memory of the computer. The measurement error of the angle and the light intensity caused by the experimental setup are of 2×10^{-5} rad and 0.1%, respectively. It should be

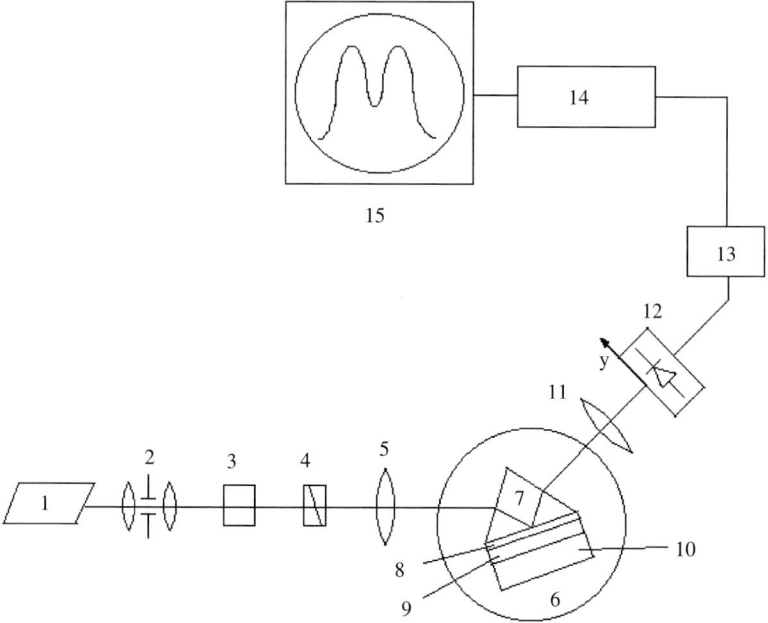

Figure 4.5. Experimental setup for measurement of waveguide parameters: light source (1), collimator (2), attenuator (3), polarizer (4), lenses (5,11), rotary table (6), prism coupler (7), gap (8), waveguide (9), substrate (10), CCD-array detector (12), intensity measuring device (13), analog–digital converter (14), PC (15).

noted that the light beam propagates through lens (11), whose focus plane coincides with the photodetector location. The relative position of the measuring prism (7), lenses (5) and (11) is chosen to satisfy the following conditions: the back focal length of the lens (5) should not be less than the distance between the principal plane of this lens and the point of the input of radiation into the waveguide. Furthermore, the front focal length of the lens (11) according to the conclusions of Section 3.2 should obviously be greater than the distance between the output face of the prism coupler and the principal plane of the lens (11). This experimental design provides the recording of the angular Fourier spectrum (intensity spectrum) of the reflected light beam when the guided modes is excited by the prism coupler. We will call such distribution as *the angular Fourier spectrum of a mode* and the measurement technique as *the spatial Fourier spectroscopy of guided modes*. The typical Fourier spectrum of the guided mode (in the direction normal to the observed *m*-line; see Figure 3.1, section A–A) is depicted in Figure 4.6.

The angle φ is counted off from the normal to the output face of the prism coupler (see Figure 4.7). The recorded parameter is the power of the

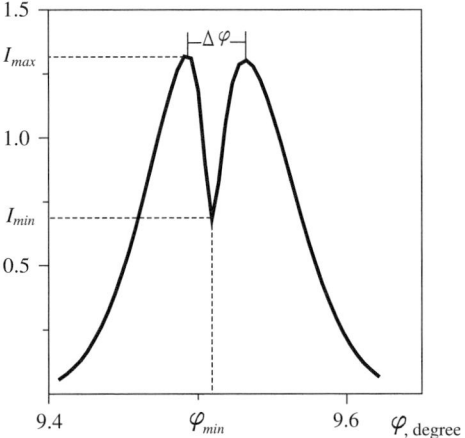

Figure 4.6. Distribution of intensity over cross section of the reflected light beam in case of the guided mode excitation by the prism coupler.

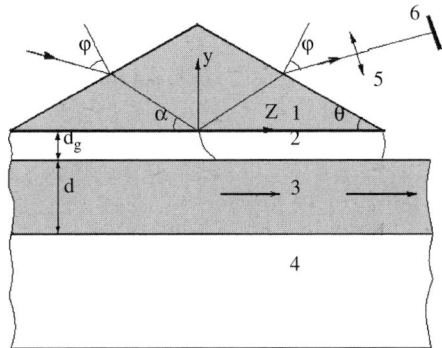

Figure 4.7. Excitation of a guided mode by the prism coupler: prism (1), gap (2), thin film (3), substrate (4), lens (5), photodetector (6).

reflected light at some fixed incidence angle φ_j. The function $I(\varphi)$ is recorded experimentally at discrete points φ_j ($j = 1, 2, \ldots, m$).

While measuring the spatial distribution of the reflected radiation intensity one can face the negative influence of noise. Therefore, the processing of the recorded Fourier spectra of modes assumes the application of mathematical optimization methods. While finding the parameters of the extrema of the function determined experimentally in discrete points the approach based on interpolation, which is realized with the help of the normal regressive analysis, is also considered. In this case the function $I(\varphi)$ is defined as [168]

$$I(\varphi) = \sum_{i=0}^{N} c_i Q_i(\varphi), \quad c_i = \sum_{j=1}^{m} I_j Q_i(\varphi_j),$$

where $Q_i(\varphi)$ are polynomials, which are orthonormal over the system of functions φ_j, I_j the experimental value for $I(\varphi_j)$ and N is determined using the Fisher criterion [167].

It is worth noting that such calculations reduce the average statistical error of the minimum coordinate determination in $m^{1/2}$ times. The accuracy determined for the mode excitation angle is approximately equal to 10^{-4} deg.

By using the previously measured parameters I_{max}, I_{min} and $\Delta \varphi$ from the recorded Fourier spectra of guided modes the real h' and imaginary h'' parts of mode propagation constants are determined by the following expressions [161]:

$$h' = n_p k_0 \sin\left[\theta_p - \arcsin\left(\frac{n_a}{n_p}\sin\varphi_{min}\right)\right] - \frac{\delta^2 - 1}{\delta^2 + 1}|\Delta h|, \quad (4.2.1)$$

$$h'' = P\frac{1+\delta^2}{8\delta}|\Delta h|, \quad (4.2.2)$$

where

$$|\Delta h| = \frac{La}{2\sqrt{A}}, \quad (4.2.3)$$

$$B = \left(S + \frac{P}{4S}\right)^2, \quad \delta = \sqrt{\frac{n_s^2 - n_a^2}{n_p^2 - n_s^2}}, \quad S = \frac{2\delta}{1+\delta^2}, \quad (4.2.4)$$

$$L = \cos\varphi_{min}\left(\cos\theta_p + \frac{\sin\theta_p \sin\varphi_{min}}{\sqrt{n_p^2 - \sin^2\varphi_{min}}}\right).$$

We can find the parameters A, B and P from the following expressions:

$$A = y\left\{\frac{P}{2}(1+y_1) - B^2 + \sqrt{\left[\frac{P}{2}(1+y_1)\right] - BPy_1}\right\}, \quad (4.2.5)$$

$$I_{max} = I_0 \exp\left(-\frac{1}{y_1}\right)\left(1 - \frac{P}{A/4y_1 + B}\right), \quad (4.2.6)$$

$$I_{\min} = I_0\left(1 - \frac{P}{B^2}\right). \tag{4.2.7}$$

Here $y_1 = (a/\Delta\varphi)^2$, $\varphi_{\min}$ is the angular coordinate of the minimum in the intensity distribution, $I_{\min}$ the intensity minimum $I_{\min} = I(\varphi_{\min})$ of the recorded distribution $I(\varphi)$, k_0 the free-space wavenumber, and $I_{\max}$ is as defined in Figure 4.6. θ_p is the angle with the base of the measuring prism.

The $\varphi_{\max}$ value is related to the angular coordinate of the point in the distribution $I(\varphi)$, where $I = I_{\max}$.

The function Δh used in expression (4.2.3) determines the influence of the prism coupler on the measured value of the propagation constant of the guided mode. The utilization of the theoretical model for describing the process of the light reflection from the prism coupler that is in optical contact with the guiding structure gives the possibility [161] to determine Δh, which depends on the optical parameters of the dielectric gap and on its thickness, i.e. $\bar{h} = h + \Delta h$. Here, $\bar{h}$ and h are the mode propagation constant in the presence and absence of the prism, respectively.

This approach can be explained by considering the process of the guided mode excitation by the prism coupler described in Ref. [160]. After passing limited beam through the lens system the intensity distribution of the beam reflected from the prism coupler (Figure 4.7) can be described in the paraxial region of the focal plane of the lens as

$$I(x, y) = I_0 |R(\theta, \varphi)|^2 |\Psi(\theta, \varphi)|^2,$$

where R is the reflection coefficient of the prism, Ψ the Fourier transform of the incident beam and κ the coefficient depending on the system magnification, $\theta = \kappa x$, and $\varphi = \kappa y$.

After the application of the CCD array the recorded one-dimensional distribution of intensity will be described by the expression

$$I(y) = I_0 |R(\varphi)|^2 \Psi(\varphi)^2. \tag{4.2.8}$$

In the case of the excitation of the waveguide by the Gaussian beam, Fourier transform is described by the expression

$$\Psi(\varphi) = \exp[-(\Delta\varphi/a)^2],$$

and the realization of the conditions of the weak coupling between the prism coupler and the waveguide as

$$\exp\left(-2\sqrt{(h'^2 - k_0 n_g^2)}d_g\right) << 1 \tag{4.2.9}$$

the reflection coefficient can be written as

$$|R|^2 = 1 - \frac{P}{A^2 + B^2}, \qquad (4.2.10)$$

where d_g is the gap thickness, a is the beamwidth and parameters A, B, P are determined by expressions (4.2.4)–(4.2.5).

At $\varphi = \varphi_{\min}$ we have

$$P = 4S \frac{h''}{\Delta h},$$

$$\delta = (n_p^2 n_g^{-2})^T \sqrt{\frac{k_0^{-2} h'^2 - n_a^2}{n_p^2 - k_0^{-2} h'^2}},$$

and for the most of the optical waveguides, h' can be replaced by $k_0 n_s$ in the above expression with high accuracy. The parameter $T = 0$ corresponds to TE and $T = 1$ to TM waves. Taking into account these expressions and also (4.2.10) one can obtain expressions (4.2.1)–(4.2.3).

Thereby so, we can get the values of the mode propagation constants for the waveguide structure situated in free space, although the values of h' and h'' are measured by the prism coupler technique. Further, we will see that this method allows one to determine correctly the parameters of thin-film waveguides.

4.2.2. Determination of Thin-Film Parameters

Let us consider in detail the process of measuring the parameters of a waveguiding film using by taking as an example of the planar waveguides (Table 4.1) obtained by RF sputtering of aluminum oxide (a line 1), and of quartz glasses KV (line 2) on substrates made from quartz glass. Thin-film waveguides were made in the atmosphere of argon and oxygen (20 vol%) mixing at a pressure of 0.1 Pa. As already mentioned above, the described approach allows one to measure the mode propagation constants for waveguides with any refractive index profiles. The data for gradient waveguides,

Table 4.1. Experimentally measured parameters of the intensity distribution while determining the mode propagation constants

$\varphi_{\min}$ (deg)	$I_{\max}$ (a.u.)	$I_{\min}$ (a.u.)	$\Delta \varphi \times 10^{-3}$ (deg)
7.3031	1500	320	7.884
6.8381	2875	1090	6.631
5.1672	2960	1315	6.140

obtained in optical glass LK6 by the out-diffusion technology [123], are given in Table 4.1 (line 3).

Using the setup described above it is possible to measure the Fourier spectra of guided modes (see Figure 4.6). Thus, it is obvious that at the beginning it is necessary to determine the angular position of the normal to the target face of the prism coupler. Also when using the technique of the light beam autocollimation it is necessary to provide the normal position of the prism faces and the working plane of the matrix photodetector to the axis of the incident beam. The processing of the recorded spectra with the help of advanced mathematical methods allows one to find the improved values of the extrema of the recorded distribution I_{max}, I_{min}, φ_{max}, φ_{min} and also $\Delta\varphi = \varphi_{max} - \varphi_{min}$ (see Table 4.1). It is necessary to emphasize that the angles φ_{max}, φ_{min} are to be measured at the greatest possible contrast $K = (I_{max} - I_{min})(I_{max} + I_{min})$ of the distribution, which is achieved by the variation of the gap thickness.

The obtained data processed according to the algorithm, stated in Section 4.2.1, give the possibility to determine the values of the real h' and imaginary h'' parts of the mode propagation constants. These values are given in Table 4.2 (here the parameter p is related to h'' by expression (3.2.8b)).

While determining the real part of h, unlike that of the traditional waveguide measurement methods, the angular position of the resonant minimum φ_{min} of the recorded intensity distribution, its angular width $\Delta\varphi$ and the contrast of the observed picture are taken into account. The same data are taken into account while determining the imaginary part of the propagation constant. The use of the complex h values, in this case for *any* of the two modes, gives the possibility of determining the refractive index n, absorption coefficient k and film thickness d. These values are obtained from the corresponding dispersion equations [168,169]

$$\left[f_1\left(\frac{\varepsilon}{\varepsilon_s}\right)^T + f_2\left(\frac{\varepsilon}{\varepsilon_g}\right)^T\right]\cos f_3 d + i\left[f_3 + \frac{f_1 f_2}{f_3}\left(\frac{\varepsilon^2}{\varepsilon_1\varepsilon_2}\right)^T\right]\sin f_3 d = 0, \tag{4.2.11}$$

Table 4.2. Mode parameters determined from results given in Table 4.1

$k_0^{-1}h'$	p (dB/cm)	$(k_0^{-1}h')^*$	p^* (dB/cm)
1.45816	12.3	1.45815	12.2
1.46254	9.4	1.46253	8.8
1.47828	1.8	1.47828	2.0

*The h' values are obtained using the technique described in Ref. [24], and p values are obtained using the method given in Section 2.5. Considered.

where

$$f_1 = \sqrt{k_0^2 \varepsilon_s - h^2}, \quad f_2 = \sqrt{k_0^2 \varepsilon_g - h^2}, \quad f_3 = \sqrt{k_0^2 \varepsilon - h^2},$$

$$\varepsilon = \varepsilon' + i\varepsilon'', \quad \varepsilon' = n^2 - k^2, \quad \varepsilon'' = 2nk.$$

Here, $T = 0$ for waves of TE and $T = 1$ for TM polarization.

According to the algorithm of solution of dispersion equations in a complex plane, proposed by my co-workers [126] we do not need to know the mode number the value of which is usually required for solving the dispersion equation. The experimenters can appreciate the advantages of this method because some situations arise while investigating the multi-mode waveguide when it is impossible to provide the excitation conditions for all modes. And there are situations while measuring the thin film parameters when $n > n_p$ (but $h'/k_0 < n_p$). Such situations frequently take place in the study of semiconductor films.

It is known that in case of mode excitation in thin-film structures by a prism coupler the results of the traditional integrated-optics measurement methods depends on the gap thickness (i.e. the coupling degree). The gap thickness is specified by applying a mechanical pressure to the thin-film structure at its clamping to the base of the prism coupler. Despite the fact that during measurements this pressure is usually kept constant, it is impossible to keep the identical gap thickness for a series of measurements. This is caused by the difference in the quality of the thin-film structure surfaces, and by its surface finish when it was prepared for measurements. In order to avoid the influence of the prism coupler on the measurement results a number of approaches and techniques have been proposed [82, 170]. However, in the measurement technique considered in this chapter this problem was solved. The algorithm developed for determining h takes into account the influence of the prism coupler (see expressions (4.2.1)–(4.2.5) and [161]). Thus, the measurement result does not depend on the conditions of the mode excitation, even if the Fourier spectrum of the guided mode is changed. The angular dependencies of the reflected radiation intensity $b(\varphi)$ recorded at a different gap thickness d_g and the values of the corresponding optical losses are depicted in Figure 4.8. The accounting of the influence of the prism coupler raises the possibility of the increase in the accuracy of determining the parameters of the guided mode. The systematic error of measurements of the refractive index δn is of 10^{-6}, of the optical losses $\delta p/p$, and of the film thickness $\delta d/d \sim 0.3\%$.

However, direct experimental confirmation of this fact causes some difficulties, as it requires the determination of the mode propagation constants by an independent method that provides higher accuracy than the accuracy

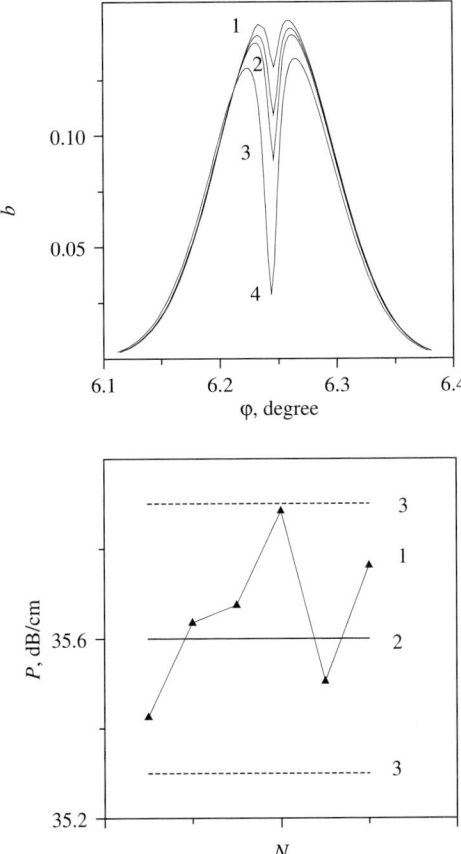

Figure 4.8. Changes in Fourier spectra at the gradual decrease of d_g (a): curves 1–4 respectively, and (b): optical losses measured at different conditions of the guided mode excitation by the spatial Fourier spectroscopy (curve 1) and by the track scanning method (curve 2) with the measurement error (curve 3).

of the prism-coupling technique. But no such reliable methods have been developed to date. As such, some attempts have been made to get indirect confirmations. In experimental measurements of the mode propagation constant the value of h'/k_0, where k_0 is the free-space wavenumber, is sometimes less than the substrate refractive index. This contradicts the physical explanation. Thereupon, we will consider the results of the following experiment.

The thin-film waveguide was fabricated by RF sputtering of aluminum oxide target on the substrate made of optical glass TK14 ($n_s = 1.61096$).

This waveguide has the thickness of 4.1 μm, and supports four optical modes of TE polarization and have the highest mode with a frequency close to the cutoff one. In order to maintain the identical conditions of measurements the recording of Fourier spectra for different modes is done at the constant clamp of the prism to the guiding structure. The values of h' for TE modes, determined by the spatial Fourier spectroscopy method, and h'^*, measured according to the method described in Ref. [24], are represented in Table 4.3. The substitution of any two of the obtained values of h' into the dispersion equations for a planar waveguide allows one to determine the refractive index and the waveguide thickness. Their averaging gives $\langle n^* \rangle = 1.63284, \langle d^* \rangle = 4.07$ μm, $\langle n \rangle = 1.63280, \langle d \rangle = 4.13$ μm. In this case, the waveguide with the values n^* and d^* will be the three-mode waveguide and with n and d should be the four-mode waveguide. The last fact is experimentally observed. Therefore, the given data indicate the higher accuracy of this technique while measuring the real part the mode propagation constant.

To check the feasibility of the obtained data the thin-film parameters can be determined by the known and independent methods. Measurements of optical losses are performed by the method of the fiber scanning along the waveguide [93] or by the photometric measurement method (measurement error is 0.3 dB/cm, see Section 2.5).

The film thickness is measured by the interferometer MI-11 (error is about 0.01 μm) or the mechanical stylus with accuracy of 0.02 μm. The measurement results for the waveguide fabricated by RF sputtering of quartz glass on substrates made from the same quartz glass in the atmosphere of argon and oxygen (5:1) are shown in Table 4.4. In this case h' was determined experimentally with an accuracy of 2×10^{-5}. This was caused by the error in the φ_{min} determination by the goniometer. The measurement error of losses was of 0.2 dB/cm. It is evident from the analysis of the results, stated in Table 4.4, that the obtained values of the waveguide parameters satisfactorily correspond to the data obtained by the independent methods.

It is possible to illustrate the reproducibility of the results obtained by the given method during the realization of a series of consecutive measurements

Table 4.3. Parameters of the thin-film waveguide modes

Mode number	h'/k_0^*	h'/k_0
0	1.63134	1.63134
1	1.62683	1.62699
2	1.61971	1.61990
3	1.61092	1.61110

Table 4.4. Parameters of SiO_x–SiO_2 structure ($n_s = 1.45670$, $n_c = 1.0003$)

m	h'/k_0	P (dB/cm)	h'/k_0^*	P^* (dB/cm)	d (μm)	n	k
0	1.47828	12.2	1.47828	12.2	1.86		
1	1.46254	9.02	1.46253	8.8		1.47973	1.26×10^{-5}
2	1.45816	2.2	1.45815	2.0	1.86*		

*Results obtained by independent methods.

Table 4.5. The stability of the mode parameter measurement results

Entry	$h''k_0^{-1}$ ($\times 10^{-6}$)	$\langle h''k_0^{-1} \rangle$ ($\times 10^{-6}$)	$h'k_0^{-1}$	$\langle h'k_0^{-1} \rangle$
1	9.92		1.472542	
2	10.03		1.472546	
3	10.20		1.472543	
4	10.09	10.0 ± 0.25	1.472547	1.472544
5	9.97		1.472542	$\pm 6 \times 10^{-6}$
6	9.81		1.472547	
7	10.10		1.472542	

of parameters of the thin-film waveguide described above. The measurements are made in a series after a complete repositioning of the experimental setup. The results are presented in Table 4.5, where the values for the separate measurement of $h''k_0^{-1}$ and an average value $\langle h''k_0^{-1} \rangle$ for the whole series are given.

The waveguide losses measured by the independent method are of 8.8 dB/cm or in recalculation on $h''k_0^{-1}$ are $10.2 \times 10^{-6} \pm 0.4 \times 10^{-6}$. Here, it should be noted that the given data may serve as an experimental estimation of errors of the measurement method. According to the normative recommendations this error a in practice is estimated as a casual error on a set of devices of the given type or with the iterative measurements made after complete repositioning of the single setup.

Unfortunately, the problem becomes more complicated while investigating the properties of thin-film waveguides with small optical losses. In this case there is a problem in accounting for the leakage of the propagated light energy from the edge of the prism coupler. Since the total attenuation of the light propagating along the film is measured in the considered configuration (see Figure 4.1), at a small value of optical losses the mode track goes beyond the bounds of the measuring prism and this "increases" the determined absorption coefficient. The optical losses, measured during the gradual increase of the clamp of the prism coupler to the examined structure, are depicted in Figure 4.9. As it is evident from the analysis of the data the values of measured losses at the definite clamp (when the gap thickness d_g is

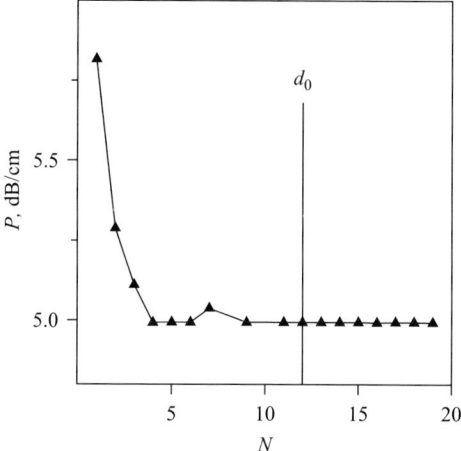

Figure 4.9. Dependence of measured optical losses on the clamp of the prism coupler to the structure under test.

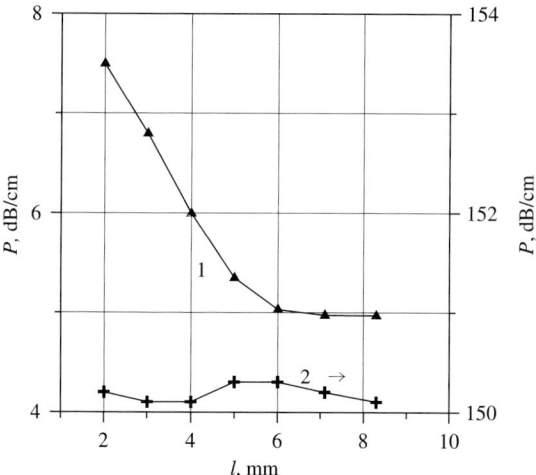

Figure 4.10. Dependence of measured optical losses on the distance between the light input point and the prism edge.

smaller than d_0) become stabilized. In the figure, the larger serial number of measurements N corresponds to the smaller values of the gap thickness.

A similar picture is observed while scanning the input point of the light beam over the prism base (Figure 4.10). The analysis of the experimental data shows that one needs to keep regular experimental conditions while measuring the absorption coefficient at $k < 10^{-5}$ in order to obtain the correct result. There are no such difficulties in measuring the greater values of k (see Figure 4.9, curve 2).

The problem of determining small optical losses can be solved by preventing the "leaking" of the light energy from the prism edge. One needs to make the gap thickness as small as possible to increase the efficiency of the light coupling from the guiding film into the prism. It is not difficult to realize this practice, but there are problems with the applicability of the stated method in the case of the weak coupling between the prism and the waveguide (see expression 4.2.9). Besides, it is difficult to keep the gap parallel between the base of the prism coupler and the waveguide due to the roughness of their surfaces and the deformation of the substrate because of large mechanical pressure. All these factors lead to additional errors while determining the optical waveguide losses $< 5\,\text{dB/cm}$. The error ranges 5–20% [170].

Hence, the considered approach allows one to measure the parameters of the guided mode, and to determine the refractive index, the absorption coefficient of film and its thickness at the same time.

Chapter 5
Characterization of Thin Films by Prism Coupling of Leaky Modes

5.1. Basic Concepts and Instrumentation 98
5.2. Determinations of Waveguiding Film Parameters 103
5.3. Leaky Modes in Thin-Film Structures 106
5.4. Determination of Parameters of Metal Films and Surface Layers of Bulk Metal by the Plasmon Modes Excitation Technique 110

Wide applications of thin films in optics and electronics and the profound integration of thin-film elements in microelectronics are required to improve the existing techniques of measuring thin-film parameters and to develop new ones. The changes in strategy of controlling the fabrication of thin-film structures in microelectronics require the fabrication of test samples together with the final product. These samples are reserved from the process of manufacturing in specified stages, and then their parameters are measured. Thus the contact waveguide methods can be applied for testing and for measuring the parameters of such structures.

There is often the problem in optics and electronics to determine the parameters of thin films in the case when their refractive index is lower than the substrate index. In this chapter, we will consider questions of determining the refractive index, absorption coefficient, and thickness of such thin-film structures. The possibility to determine the refractive index and the film thickness in such structures of when the leaky modes in are excited, has already been illustrated in Refs. [35–37]. The application of the spatial Fourier spectroscopy method for the determining parameters of such films is difficult because on observing the Fourier spectra of the reflected light beam during the excitation of leaky modes (as well as plasmon modes), the recorded distribution was found to be a non-symmetrical function. This causes some difficulties in the processing and interpretation of the recorded data. Besides, it is well known that when a surface is illuminated by laser radiation speckles spots are found in the reflected light. The size d_s of the speckle spots is estimated from the following formula that takes into account diffraction effects:

$$d_s \approx 2.44 \lambda f / D, \qquad (5.1.1)$$

where λ is the light wavelength, and f and D are the focus distance and the diameter of the lens forming the spots, respectively.

Since the surface is not ideally flat, every point of the illuminated surface scatters the light in the angle range that is equal to $\sim D/f$. The radiation scattered from different points of the surface interferes and this forms the accidental interference picture in the form of the speckle picture. In optical systems the aperture is usually changed in the range from $f/1.0$ to $f/50$, and hence, the typical size of speckles in the observation plane varies in the range from 2 to 100 μm [171]. This is clearly recorded when light beam is focussed into the small spot. In the case of the leaky mode excitation we have to use exactly this size of light beams while recording the Fourier spectra. In this situation the recorded distributions are so noisy that the measurement of film parameters become impossible. Thus, attempt to modify the integrated-optical methods to conform the characterization of structures such as SiO_2–Si, which are of practical interest, have been made.

The results stated in this chapter are the improvements of waveguide measurement methods when the leaky mode is excited in absorbing thin films by the prism-coupling technique. This approach is based on the recording of the angular dependence of the reflection coefficient of the light beam reflected from a prism coupler in case of a tunnel excitation of optical modes in thin-film structures or surface layers.

5.1. Basic Concepts and Instrumentation

In the case of the excitation of guided modes by the traditional scheme (Figure 3.1) one can observe series of dark m-lines in the reflected light (Figure 5.1). The typical structure of the corresponding angular dependence of the

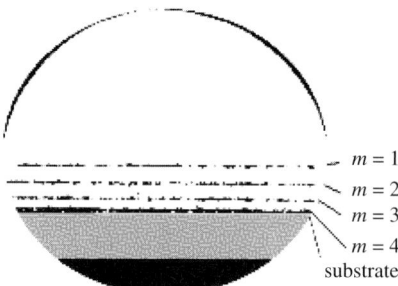

Fig. 5.1. Observed spectrum of guided modes in the case of the excitation of the waveguiding structure by the converging light beam.

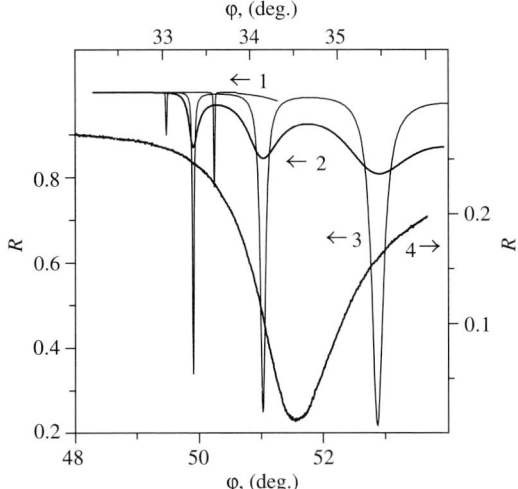

Fig. 5.2. Angular dependence of the reflection coefficient of the light beam in the case of the excitation of guided modes in structure SiO_x–SiO_2 (curve 1), leaky modes in structures "SiO_x–Si" and "SiO_x-optical glass K8" (curves 2 and 3, respectively) and plasmon mode at the Al-film interface (curve 4).

reflection coefficient of the light beam reflected from the prism coupler in the case of the excitation of guided modes is depicted in Figure 5.2 (curve 1).

A similar dependence can be recorded in case of the excitation of leaky modes in non-waveguiding structures (Figure 5.2, curves 2 and 3), and of plasmon modes at the interfaces of metal films (curve 4). It turned out that the appropriate processing of such reflection coefficient allows one to restore the real and imaginary parts of the propagation constants of the leaky mode.

After an accounting of the description of the process of the mode excitation in thin-film structures stated in Ref. [172], we will explain the approach considered above. Let us consider the thin-film-directing optical modes, which is deposited on the substrate with permittivity ε_s. The thin-film structure is in contact with an isosceles prism, as shown in Figure 3.19. The prism with the angle θ at the base, the medium surrounding the prism, and the gap with thickness d_g have real permittivity ε_p, ε_a and ε_g, respectively, and $\varepsilon_p > \varepsilon_g \geqslant \varepsilon_a$. The structure is excited by the Gaussian beam with the width a_w, whose axis is at the angle φ with the normal to the input plane of the prism (see Figure 4.7). The energy reflection coefficient of the light is determined by the expression

$$R(\varphi) = A(\varphi)r(\varphi), \qquad (5.1.2)$$

where $r(\varphi)$ is the coefficient of the light reflection from the prism base, $A(\varphi) = 16\kappa^2/(1+\kappa)^4$, $\kappa = (\varepsilon_p/\varepsilon_a)^T \cos\varphi/\sqrt{\varepsilon_p/\varepsilon_a - \sin^2\varphi}$, $T = 0$ for TE waves and $T = 1$ for waves with TM polarization [168].

It is well known that the dependence of $r(\varphi)$ has one minimum coordinate at the excitation of the guided mode; we will denote it has φ_0 and $r(\varphi_0) = r_0$. The complex propagation constant h can be determined from the experimental dependence $r(\varphi)$. If $r(\varphi)$ is measured in the angle range $\varphi_0 - a \leqslant \varphi \leqslant \varphi_0 + a$, one can get the following expression for the propagation constant h:

$$h = \beta + \frac{\sin\alpha}{w}\{P_4^{(0)} + i[P_1 - P_2(1-\delta)^2(2\delta)^{-1}]\}, \tag{5.1.3}$$

where $k_0 = 2\pi/\lambda_0$ is the free-space wavenumber,

$$\delta = (\varepsilon_p \varepsilon_g^{-1})^T \sqrt{(k_0^2 \varepsilon_g - \beta^2)(k_0^2 \varepsilon_p - \beta^2)},$$
$$w = w_0(\cos\varphi_0)^{-1}\sqrt{1 - \varepsilon_a \varepsilon_p^{-1} \sin^2\varphi_0},$$
$$\beta = k_0\sqrt{\varepsilon_p}\cos\alpha,$$
$$\alpha = 0.5\pi - \theta + \arcsin\left(\sqrt{\varepsilon_a \varepsilon_p^{-1}}\sin\varphi_0\right).$$

The δ value is calculated under the condition $\operatorname{Re} h = \beta$, and the magnitudes β, α, w are determined at $\varphi = \varphi_0$. The parameter P_1 can be found from the equation

$$[G(-P_1)]^{-1}\operatorname{Re}\int_0^{a_1} G\,dP_4 = \left(2a - \int_{\varphi_0-a}^{\varphi_0+a} r\,d\varphi\right) 0.5 k_0 \sqrt{\varepsilon_a} w_0 (1-r_0)^{-1}, \tag{5.1.4}$$

where

$$G(t) = \frac{i}{\sqrt{2}}\int_{-\infty}^{\infty} \frac{\exp(-\tau^2)}{it - \tau\sqrt{2}}\,d\tau,$$
$$t = -P_1 + iP_4,$$
$$a_1 = k_0\sqrt{\varepsilon_a}w_0 a.$$

The values σ, $p_4^{(0)}$ and $p_2 = |p_2|\exp(i\sigma) = -2iw\Delta h\delta(1-\delta)^{-2}$ can be obtained from the following expressions:

$$2|p_2| = -P_1 \pm \sqrt{P_1^2 + P_1\sqrt{0.5\pi}(1-r_0)[G(-P_1)]^{-1}}, \tag{5.1.5}$$

$$\sigma = k_0\sqrt{\varepsilon_a}w_0\left(\int_{\varphi_0}^{\varphi_0+a} r\,d\varphi - \int_{\varphi_0-a}^{\varphi_0} r\,d\varphi\right)N^{-1}, \quad (5.1.6)$$

$$p_4^{(0)} = \sigma P, \quad (5.1.7)$$

where ρ is equal to 0 or 1,

$$P = [p_1 G(-p_1) + \sqrt{0.5\pi}]\{(1+|p_2|p_1^{-1})[p_1\sqrt{0.5\pi} + (1+p_1^2)G(-p_1)]\}^{-1},$$

$$N = 8|p_2|\sqrt{\frac{2}{\pi}}\left\{\mathrm{Im}\int_0^{a_1} G\,dp_4 - \left(\frac{|p_2|}{p_1}+1\right)[ReG(-p_1+ia_1) - G(-p_1)]P\right\}.$$

Expression (5.1.3) allows one to determine the complex propagation constant of modes for the layer with an arbitrary refractive index profile. The influence of the prism coupler on the measured results is also taken into account.

The automated device is used to perform such measurements. The principal scheme of this setup is shown in Figure 5.3. The power of the light beam reflected from the prism coupler at the given incidence angle is the recorded parameter [173]. The He–Ne laser with the wavelength 632.6 nm is used as the radiation source. The light beam incident onto the prism coupler (7) is positioned on the rotary table (10). In this case the prism is made from optical glass TF12, with the refractive index equal to 1.77905 at the wavelength of 0.6328 μm.

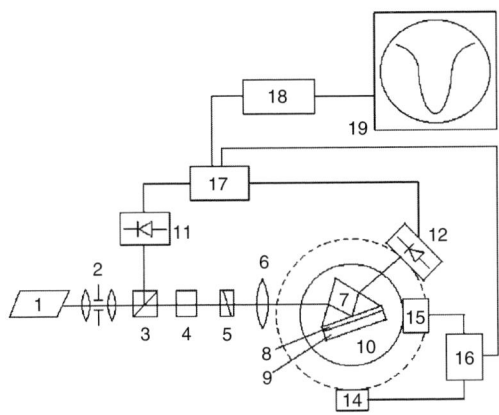

Fig. 5.3. Experimental setup for measuring the angular dependencies of the reflection coefficient: light source (1), collimator (2), beam splitter (3), attenuator (4), polarizer (5), lens (6), prism coupler (7), gap (8), thin-film structure (9), rotary table (10), photodetectors (11, 12), stepping motors (14, 15), block of synchronous operating of stepping motors (16), block of channel comparison (17), analog–digital transformer (18), computer (19).

Measurements can be made for different light polarizations as well. Device construction allows one to change the light beamwidth in the range 70–500 μm. The beamwidth is measured by the intensity level $I = I_0 e^{-1}$. The investigated sample is pressed to the measurement prism in such a way that it provides optimal conditions for the excitation of optical modes in thin-film structures. The complex-mode propagation constant h depends on the optical and the geometrical parameters of the film and the gap (8). It also depends on parameters of the surrounding medium, which should be kept constant. The angle between the incident light beam and the prism plane is changed with the help of the stepping motor (15). The discretization step of the rotation angle of the table is equal to 20 (angular seconds). The angular dependence of the reflection coefficient is recorded with the help of the photodetector (12), being synchronously moved by a second analogous stepping motor (14). It is appropriate to use the photodiode with large area. The unit controlling the stepping motors (16) is synchronized with the channel comparator (17). After digitizing (12-digit analog–digital converter) the signal is transmitted to the computer online—storage in the direct access mode.

The computer program based on theory [172] with the final expressions stated above allows one to process the recorded distribution of the reflection coefficient and to find the real and imaginary parts of the guided mode propagation constant. The real h' and imaginary h'' parts of the propagation constant are given by the angular position of $\varphi_{\min}$ (Figure 5.4), as well as in the case of Fourier spectra processing, also by the contrast of picture $I_{\min}/I_0$

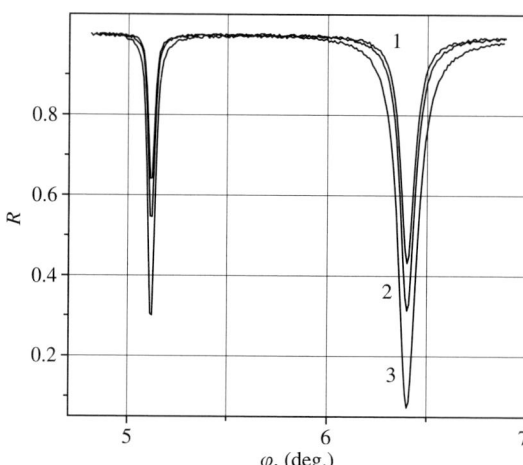

Fig. 5.4. Angular dependencies of the reflection coefficient of the light beam reflected from the prism coupler at the excitation of one waveguide at different coupling power ($g_1 < g_2 < g_3$).

and by the angular width of the resonant minimum. This gives the possibility to determine at the same time the refractive index, thickness, and the absorption coefficient k of the film material [168]. In its turn, k is related to the high-frequency conductivity of the material $\sigma(\omega)$. If the mechanism of the conductivity is quite clear it is not difficult to find σ_c for the direct current [60]:

$$\sigma(\omega) = \sigma_c/(1 + \omega^2\tau^2), \tag{5.1.8}$$

where τ is the relaxation time.

Hence, we have the method of measuring thin-film parameters that allows one to perform a comprehensive analysis of films.

5.2. Determinations of Waveguiding Film Parameters

Before considering the question of determining the parameters of thin-film when the leaky modes are excited in them, let us demonstrate the possibilities of this technique on the example of thin-film waveguides. These are the structures, which have the refractive index greater than the refractive index of the substrate. Let us consider the waveguiding structures obtained by RF sputtering of quartz glass K8 in the atmosphere of argon and oxygen (4:1) on the substrate made from the same glass [174]. Previously, the substrates were deeply mechanically polished. Such waveguide can direct three modes of TE polarization.

We can note the following basic phases of the measurement of the angular dependence of the beam reflection coefficient during the excitation of such thin-film structures. As the measurement device works automatically we have to define the reference point for the determination of the angle. It can be easily done if one chooses the angular position of the photodetector (12) during the recording of the directly-propagated light beam as a reference point. At this moment the prism moves out of the beam zone. Now it is not difficult to determine the angular position of the normal to one of the prism faces. Taking into account the geometry of the prism coupler and optical parameters of the substrate and the prism material the processor previously determines the range of measured angles and gives commands to the stepping motors to prepare the system for metering. Then, for the given set of samplings and number of measurements, the automated collection of statistical data, their averaging, and the recording of distribution into the computer memory take place for each sampling. The processing of such distribution allows one to determine the values of guided mode propagation constants and then these values are used to calculate the parameters of thin-film structures. The examples of such measurements and calculation results are given in Table 5.1. The accidental error in determining the propagation constant of the real part is 1×10^{-5}, and the imaginary part is about 2%.

To check the feasibility of the obtained results, the thin-film parameters can be determined by independent methods. The measurements of the optical losses for the second mode by the method of scanning of the fiber along the waveguide [93] (the measurement error is equal to 0.2 dB/cm) gave the values of optical losses of about 5.5 dB/cm. Film thickness measured with the mechanical stylus with the accuracy of 0.02 μm was equal to 2.50 μm. As we can see, the obtained result is in good agreement with the data given in Table 5.1.

We should mention that the method considered here can be applied for the determination optical losses of guided mode in planar waveguides with an arbitrary profile of the refractive index.

As it has been already mentioned in Chapter 4, while measuring the parameters of film by the prism-coupling technique the accuracy of the result depends on the coupling between the prism and the waveguiding structure (i.e., the thickness of the gap between the film and the prism). The angular dependence of the beam reflection coefficient will be different in practice also (Figure 5.4). However, in the case of the recording of the angular Fourier spectrum, the influence of the prism coupler is taken into account [172] as a generalized approach to the solution of the problem of the reflection of light beam from the prism coupler. This leads to the independence of the measurement results under the experimental conditions. The systematic error caused by the approximations of the used theoretical model is equal to $\sim 2 \times 10^{-6}$ for the refractive index measurement, and $\sim 1\%$ for the absorption coefficient and thickness (i.e. while measuring the

Table 5.1. Parameters of thin film determined using the values of the guided mode propagation constants

	Fourier spectroscopy technique		Recording of angular dependence of the light reflection coefficient	
	$m = 0$	$m = 1$	$m = 0$	$m = 1$
$h' k_0^{-1}$	1.46755	1.45814	1.46748	1.45810
$h'' k_0^{-1}$	9.98×10^{-6}	6.51×10^{-6}	1.02×10^{-5}	6.71×10^{-6}*
n	1.47104		1.47099	
k	1.03×10^{-5}		1.08×10^{-5}	
d (μm)	2.49		2.53	

*The value corresponds to the optical losses of 5.5 dB/cm.

parameters of a film with thickness of ~1 μm and absorption coefficient of ~10^{-5}, their errors were $\delta d = 100$ Å and $\delta k < 10^{-7}$).

One of the peculiarities of determining the thin-film parameters during the recording of the angular dependence of the reflection coefficient is the selection of the probe light beamwidth. We can obtain quite correct results only by using wide light beams. In the ideal situation this corresponds to the excitation of the guided mode by the plain wave. Distributions of the reflection coefficient at the excitation of thin-film waveguide by light beams with different beamwidths are given in Figure 5.5. The results of thin-film parameter measurements correspond to the data of independent measurements for curve 3 only.

While recording the Fourier spectra by this method there is also the problem of energy "leakage" from the prism coupler. This is owing to the utilization of the same model describing the processes of light propagation in such structures. This is also caused by the fact that in this method we determine the total attenuation of radiation in film. At the small value of optical losses the radiation leaves the limits of the measuring prism and "increases" the measured value of the absorption coefficient.

In this case, as also in the case of Fourier spectrum recording, the values of the measured losses become stable at the thickness of buffer layer less than some definite value. Determination of small losses in this configuration is possible if we prevent the light energy "leaking" out of the prism limits by decreasing the gap thickness. Unfortunately, the breaking of the gap

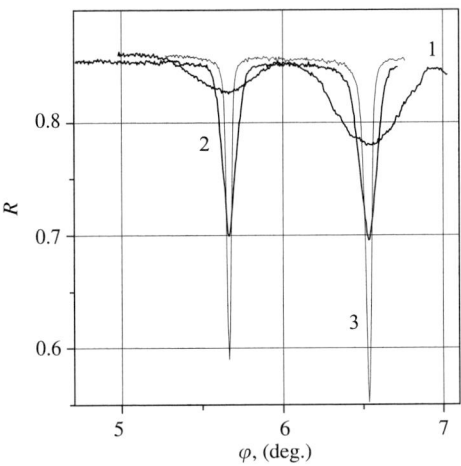

Fig. 5.5. Angular dependence of the reflection coefficient at the waveguide excitation by the light with the beamwidth equal to 100, 150 and 450 μm (curves 1, 2, 3, respectively).

Table 5.2. Results of the measurement of mode parameters with the consecutive increase in the clamping of the prism to the waveguide surface

d_g^* (μm)	$h'' k_0^{-1} \times 10^{-6}$	$\langle h'' k_0^{-1} \rangle \times 10^{-6}$
0.123	1.967	
0.131	1.423	
0.137	2.281	
0.141	1.804	
0.144	1.535	2.18 ± 0.55
0.146	1.692	
0.149	2.094	
0.152	2.372	
0.155	2.118	
0.161	2.687	
0.167	3.098	
0.172	3.014	

*d_g values are obtained by calculation.

parallelism between the base of the prism coupler and the surface of the sample under test leads to the additional error in the determination of optical losses in the waveguide. The measurement results for thin-film waveguide with optical losses ~ 2 dB/cm are given in Table 5.2. In the case considered here the measurement error of the imaginary part of the propagation constant is equal to $\sim 20\%$.

So, while measuring the absorption $\alpha < 5$ cm^{-1} (for the thin-film structure examined above, $\alpha d < 2 \times 10^{-4}$) one needs to keep some criteria of measurements in order to get the reliable result. Usually, such difficulties are not there for large values of the absorption coefficient.

5.3. Leaky Modes in Thin-Film Structures

After clarifying the principles of measurement of thin-film parameters by the scheme of prism excitation of the guided mode during the recording of the reflection coefficient are return to the question of the determination of the parameters of such thin-film structures that do not have waveguiding properties. In this section, we will consider the problems in determining the refractive index, absorption coefficient, and the thickness of non-waveguiding layer deposited on different substrates.

We can use thin films, which are also obtained by the sputtering of the quartz glass but deposited on substrates with a larger refractive index. The values of refractive indices of the substrate and the film in this case are alike.

Substrates were fabricated from the optical glass K8 and from monocrystalline silicon. The refractive indices of glass and silicon are 1.51466 and 3.510, respectively, at the wavelength 632.8 nm. These thin-film structures were obtained simultaneously during one technology cycle together with the SiO_x/SiO_2 structure considered in the previous section. Therefore thin-film structures should have quite close optical parameters and approximately equal thickness. Only the leaky modes can propagate in such structures (see Chapter 2). The angular dependence of the light beam reflection coefficient at different strengths of clamping of the sample to the prism coupler is depicted in Figure 5.6. The method of the recording of this angular distribution and the algorithm of its processing are similar to the methods applied in the case of the guided modes considered in Section 5.2. In the case of the mathematical description of properties of the reflected light beam the influence of the prism coupler on the measured parameters of the guided mode was taken into account and the film's absorption coefficient, refractive index, and thickness could be reliably determined.

Accounting the influence of the prism coupler however, leads to an ambiguous solution. In practice, one value of the propagation constant from two values obtained from the result of mathematical processing of experimental data (see expression (5.1.3)) is to be chosen. Moreover, we cannot exclude beforehand any of these two values because it can be the true value. A simple solution to this polysemantic task was found. It is enough to measure the gap thickness at two different values (i.e. at two different clamps of the sample to the prism coupler) and the true value of h remains stable, and the second

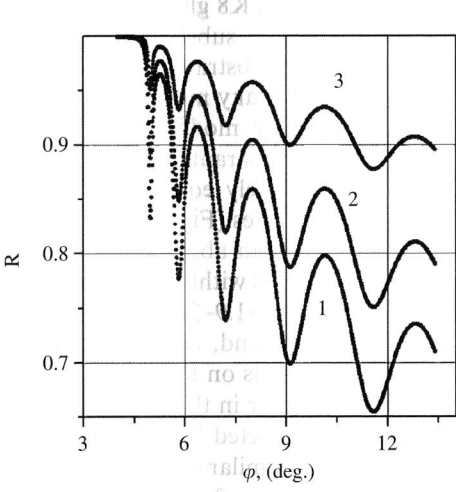

Fig. 5.6. Angular dependence of the light beam reflection coefficient at different clamping of the sample to the prism: $d_g = 0.1\,\mu m$ (1), $0.13\,\mu m$ (2) and $0.17\,\mu m$ (3).

Table 5.3. Results of processing of angular distributions depicted in Figure 5.6

Curve	h'/k_0		h''/k_0	
1	1.46512	1.46483	1.77×10^{-3}	1.22×10^{-4}
2	1.46512	1.46493	1.76×10^{-3}	7.97×10^{-5}
3	1.46512	1.46492	1.76×10^{-3}	3.39×10^{-5}

*The data are stated for mode with $m = 1$.

Table 5.4. SiO_x film parameters on different substrates

	n	$k \ (\times 10^{-5})$	d (μm)
SiO_x/SiO_2	1.47095	3.39	2.51
SiO_x/Si	1.47091	3.34	2.53
$SiO_x/K8$	1.47024	2.5	2.69

value is changed (Table 5.3). This fact clearly shows itself at the variations of the values of the imaginary part of the propagation constant [175].

If we apply the technique described above in the experiment and then process the angular distributions of the reflection coefficient (Figure 5.1), it will be possible to determine the parameters of the film deposited on the different substrates. The obtained results are given in Table 5.4.

It is difficult to explain such significant differences in the parameters of the film on the substrate made from K8 glass and the other structures by the influence of the composition of the substrate material [176]. In this case, small difference in the film and substrate refractive indices ($\Delta n = n - n_s$) causes a weak localization of the leaky mode and leads to large absorption associated with the leakage of the mode energy into the substrate. The results of the determination of the parameters of SiO_x film can confirm this fact. These films have approximately equal thickness (2.5 ± 0.03 μm) and were deposited on different substrates (Figure 5.7). As it is evident from the analysis of the given curves, the film absorption coefficient is equal to the values determined by other methods within the accuracy of 3% of the values of the substrate's refractive index ~1.9–2.0. It is obvious that the localization of the mode field in the film and, therefore, the reliability of the determination film parameters depends on the thickness of the deposited film. The dependence of the relative error in the determination of the absorption coefficient on film thickness is depicted in Figure 5.8.

All films were deposited under similar conditions and have approximately equal parameters, $n = 1.4701$ and $k = 3 \times 10^{-5}$. The errors in determination of thickness and refractive index were decreased from 6 to 0.5% and from 5×10^{-5} to 1×10^{-5}, respectively.

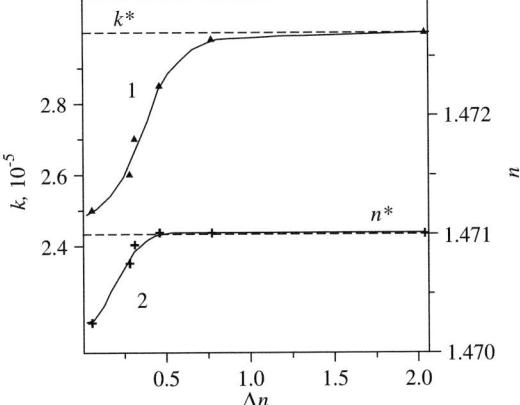

Fig. 5.7. Dependence of the measured values of the absorption coefficient (1), the refractive index (2) of films on a difference of the film refractive index and substrate index; values of k^*, n^* are determined for the SiO_x/SiO_2 structure.

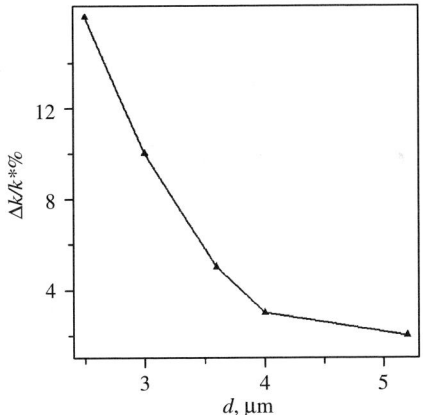

Fig. 5.8. Influence in the film thickness by the error of determination of absorption coefficient of the SiO_x film on the substrate made from K8 glass.

So, the application of this method for determining the thin-film parameters is appropriate at quite a large difference of the refractive indices of the substrate and the deposited film (at least $\Delta n > 0.5$). In the case of small Δn the application of such method is appropriate at $d \approx 5\,\mu m$.

Let us consider the question of restoration of parameters of the film with minimal possible value of k with an acceptable error in the determination of the parameters. We will use SiO_x films on silicon substrate as the sample

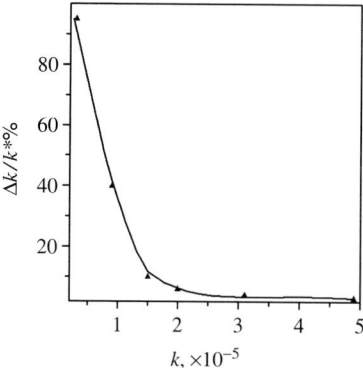

Fig. 5.9. Dependence of the determination error of SiO_x film absorption coefficient on the film thickness (Si substrate); k is determined in the case of the excitation of guided modes in structure-satellites "SiO_x/SiO_2".

under investigation. The films have almost equal thickness of about 2.5 μm. They were obtained by RF sputtering of different oxygen concentrations in the operating atmosphere. Because of different stoichiometry they had different values of the absorption coefficient. The relative error in the determination of the absorption coefficient of the film versus the film material k value is shown in Figure 5.9.

As we can see, the determination error of the absorption coefficient at values of $k < 10^{-5}$ is more than 30%. This quite large error can be explained by the fact that at k ($\sim 10^{-5}$) the measurement error of $h'' k_0^{-1}$ becomes comparable and even exceeds the k value. In this case, the determination of absorption of the light in the film becomes problematic. The determination error of the refractive index and the thickness do not exceed 5×10^{-5} and 2–3%, respectively.

Thus, the method of investigating thin-film property based on the basis of recording the light beam reflection coefficient during the excitation of guided modes can be applied for the testing and measurement of the parameters of various thin-film structures used in optics, opto- and microelectronics.

5.4. Determination of Parameters of Metal Films and Surface Layers of Bulk Metal by the Plasmon Modes Excitation Technique

The method considered above can also be applied in the case of plasmon modes propagating along the surface of metal films surrounded by dielectric media. However, to measure the thickness of metal films one needs to excite the plasmon modes at both interfaces [18]. It is obvious that due to the large absorption of visible range radiation in the metal, the excitation of the plasmon mode on the outer boundary (relative to the prism coupler) is possible only at the film thickness that is approximately equal to 300–500 A. The considered technique allows one to determine the absorption coefficient, refractive index, and the thickness of only thin metal films. The processing sequence is similar to the one described in the previous section. At film thickness >80 nm the second interface does not affect the parameters of the plasmon mode excited at the first interface of the metal film. In this case the optical parameters of thick films and surface layers of bulk metals can be determined on the basis of the recorded intensity of the spatial distribution of the reflected light beam when the plasmon mode is excited at the inner surface of the film only (Figure 5.10).

The plasmon mode propagation constant is related to the permittivity of the metal film (the surface layer of bulk metal) $\varepsilon = \varepsilon' + i\varepsilon''$ and surrounding medium ε_a by the following expression [170]:

$$h^2 = (h' + ih'')^2 = k_0^2 \frac{\varepsilon_a \varepsilon}{\varepsilon_a + \varepsilon}; \qquad (5.4.1)$$

then

$$\varepsilon' = \frac{(h'^2 - h''^2)k_0^2\varepsilon_a - (h'^2 + h''^2)^2}{z}, \qquad (5.4.2)$$

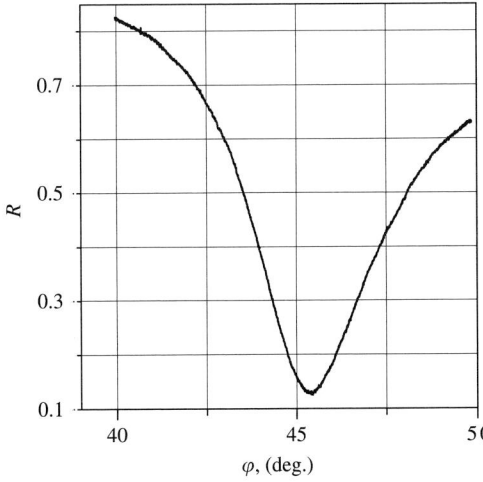

Fig. 5.10. Angular dependence of the reflected light beam intensity in the case of the plasmon mode excitation in the Al film.

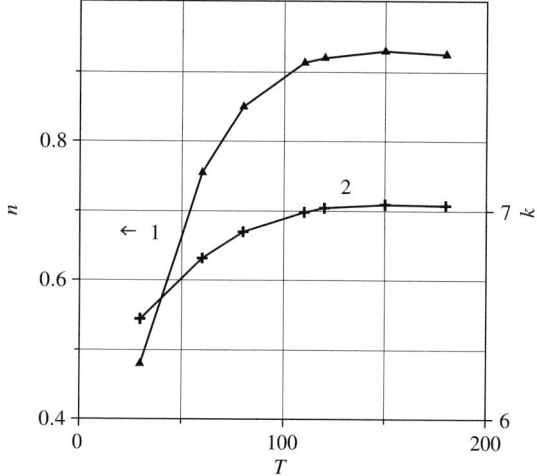

Fig. 5.11. Dependence of the refractive index n (1) and the absorption coefficient k (2) on the substrate temperature T for the aluminum film on the quartz glass substrate.

$$\varepsilon'' = \frac{2k_0^2 h' h''}{z} \varepsilon_a, \qquad (5.4.3)$$

where

$$z = (\varepsilon_a k_0^2 - h'^2 + h''^2)^2 + (2h'h'')^2.$$

By measuring the complex h, we can determine the complex permittivity of the metal and, therefore, its refractive index and absorption coefficient.

This technique of the investigation of metal property can be illustrated by the example of the measurement of the parameters of the aluminum films deposited on the quartz glass substrate. The refractive index (n) and the absorption coefficient (k) values obtained for the film were equal to 0.433 and 6.69, respectively. Errors of their determination were $\delta n/n = 0.04$ and $\delta k/k = 0.05$ [175]. As we can see the accuracy of the metal film parameter is determining similar to that of other techniques [99,100]. The results of determining parameters of aluminum films obtained by cathode sputtering at different substrate temperatures T in the deposition process are stated below. At substrate temperature $T_0 > 120°C$ the thin film with good adhesion to the substrate are obtained [173]. As is evident from the data stated in Figure 5.11 the film has a high refractive index in the temperature area $T > T_0$. This fact indicates the good quality of thin films.

Thus, we can see, the technique based on the recording of the angular dependencies of the reflection coefficient allows one to inspect the parameters of metal films and also to judge their quality.

Chapter 6
Measurements of Absorption Spectra of Thin Films

6.1. Absorption Optical Spectrophotometry: Possibilities and Limitations 114
6.2. Instrumentation of the Waveguide Spectroscopy of Thin Films 115
6.3. Special Features of Absorption Spectra Recording by Waveguide 119
Spectroscopy and their Processing . . . .
6.4. Measurement of Absorption Spectra by the Fourier Spectroscopy of 120
Guided Modes . . . .

Electrical and optical properties of semiconductors and dielectric materials are defined by the behavior of the density distribution of the electron states in the band gap. In order to understand and explain the properties of real materials one needs to determine exactly the density of such states. Although it is possible to get this information with some limitations for crystalline materials, it is much more difficult to do the same for polycrystalline and amorphous film materials. One of the reasons is that films are fabricated under non-equilibrium conditions and their state density depends on the manufacturing process. Besides, the type of chemical bond, coordinating numbers, and interatomic distance affect the state density. Altogether this as a whole does not allow one to calculate the electron state density for thin films. For this reason experimental determination is required [42]. Some information in this regard can be obtained by optical spectroscopy in the spectral range below the fundamental absorption edge. But it is difficult to measure the absorption in the thin and weakly absorbing films with the help of the existing methods. This difficulty is caused by the fact that the direct measurement of absorption when $\alpha d \ll 1$ becomes inexact [60], where d is the film thickness. Measurements of photoconductivity, which allows one to determine the material parameters when the value of absorption coefficient $\alpha \approx 1\,\text{cm}^{-1}$ for semiconductor materials, are not applicable for the study of dielectric films [177]. As such more perfect measurement methods have been developed [3].

In spite of the fact that the absorption bands are very wide in the visible range and they have feebly marked structure, complex investigations allows one to study the density distribution of the electron states in the band gap.

6.1. Absorption Optical Spectrophotometry: Possibilities and Limitations

At the time of its discovery, the spectroscopic investigations in the optical range gave the possibility to discover and analyze the atomic structure of a substance. But the enhancement of the existing methods allowed one to determine more exactly the wavelengths of the spectral lines, their profile or the shift caused by the influence of external effects.

We can mark out the interference spectroscopy among the classic methods having the maximum spectral resolution. The high resolution $\lambda/\Delta\lambda$ equal to $\sim 10^7$ allows one to achieve such an accuracy of measurement that the dependence of the spectral line width on thermal motion of radiating atoms begins to manifest itself [178]. Invention of laser has opened a new epoch in spectroscopy since it produces intensive light sources with a high degree of monochromaticity [179]. There are different methods of laser spectroscopy [129,150,180–182], which allow one to resolve the fine structure of spectral lines. By using the modern sophisticated methods and methods of classic photometry a significant number of measurements of the spectral absorption coefficient in the visible range were performed for thin-film structures. It is convenient to divide the absorption spectrum into three regions [11, 177]. The absorption coefficient α (in the first spectral range – above the optical band gap) exceeds $10^3\,\text{cm}^{-1}$. Here the light absorption is caused by "band–band" transitions. For example, the following expression can be used in order to find the optical width of the band gap E_g:

$$\alpha(\hbar\omega) = 8 \times 10^4 \frac{1}{n}(\hbar\omega - E_g)^{1/2}, \qquad (6.1.1)$$

where n the refractive index of the material studied, and $\hbar\omega$ and E_g are given in electron-volts.

Note that the quantity $\alpha(\hbar\omega)$ is related to the electron state density N by the expression

$$\alpha(\hbar\omega) = N(\hbar\omega)\frac{2\pi^2 \hbar g_e^2}{n c m_e^*}, \qquad (6.1.2)$$

where g_e and m_e^* are the charge and effective mass of the electron, respectively, and c the light speed [150].

Expressions (6.1.1) and (6.1.2) are valid for the direct transitions between the parabolic bands. The second range of the absorption spectrum lies near the fundamental absorption edge and the value of α is equal to $10-10^3\,\text{cm}^{-1}$ here. The well-known methods used for the measurement of optical absorption in both the spectrum regions give reliable results. There is also an absorption range below the fundamental absorption edge, where $\alpha < 10\,\text{cm}^{-1}$. While performing optical measurements in this spectral range some problems arise related to the feasibility of the obtained results. Straight measurement of absorption at $\alpha d \ll 1$ becomes incorrect for films with thickness equal to $\sim 1\,\mu\text{m}$ and less [60].

All this as a whole stimulates the development of more modern and perfect measurement techniques [182].

6.2. Instrumentation of the Waveguide Spectroscopy of Thin Films

The precision methods of thin-film parameter measurement considered in chapters of this book allow one to determine the refractive index and absorption coefficient of films while using a coherent source of radiation.

There is, therefore, the desire for the utilization of the above methods in order to determine the spectral absorption coefficient. The technique of measuring film absorption spectrum considered below is the development of researches in the measurement of thin-film parameters using coherent radiation in the case of non-monochromatic light. It is based on the recording of the spatial distribution of the intensity of the radiation reflected from a prism coupler when the guided mode is excited in the thin-film structure.

Devices for photometric investigations are different but all of them have common structural elements: a radiation source, a device for separation of the light beam in a narrow range of wavelength, photodetectors, a focusing, element a collimation and other optical elements. In order to process the recorded signal, it is better to connect the optoelectronic block to the computer. The device constructed for conducting such investigation has similar structural elements, and is schematically shown in Figure 6.1. The scheme is of conditional character as some elements can be more complicated or they may also be absent.

The power of the light beam, which is reflected from the prism coupler while scanning the incidence angle, is the recorded quantity. The light beam from the radiation source with controllable intensity (1) through the monochromator (3) enters into the prism coupler (7), which is the isosceles prism situated on the rotary table (10). This light beam excites the guided mode in the studied thin film (9). The complex propagation constant h of the guided mode depends on the optical parameters and on the thickness of the thin-film structure. The optical scheme allows one to perform measurements by

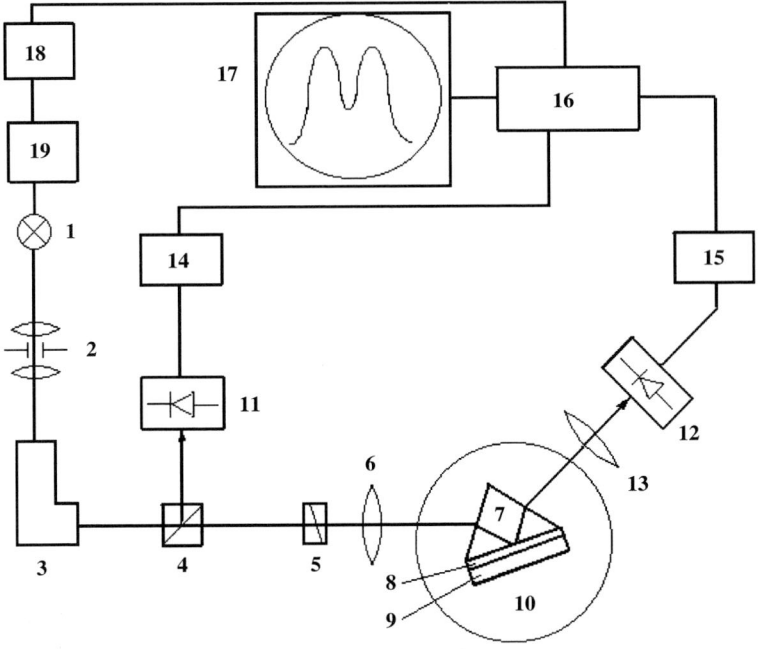

Fig. 6.1. The scheme of measuring device: radiation source with controllable intensity (1), collimator (2), monochromator (3), beam splitter (4), polarizer (5), lenses (6, 13), prism coupler (7), gap (8), investigated structure (9), rotary table (10), photodetectors (11, 12), intensity measurement devices (14, 15), analog–digital converter (16), computer (17), feedback circuit (18), power source (19).

using the radiation with a different polarization. While investigating the temperature dependence it is required to measure the temperature of the sample with sensors such as thermocouple, thermistor, etc. The possibility of the beamwidth change in the range 15 to 150 µm is laid in the device construction. The beam is supposed to have the Gaussian shape and its width is measured by the intensity level $I = I_0 e^{-1}$, where I_0 is the intensity in the center of the beam.

Lasers with variation of operating frequency or classic light sources, for example, incandescent lamps with continuous radiation spectrum or low-temperature discharging lamps, which meet the qualifying standards of the light sources in the range of the radiated wavelength, can be used as the radiation source. The incandescent lamp radiation curve is given in Ref. [183]. For the selection of a certain wavelength from the continuous radiation spectrum of the lamp the monochromators (prism or diffractive ones) are usually applied. The schematic setup for some monochromators are described in Refs. [184–186]. The monochromator can be applied in combination with any light source. After the propagation through the collimator the light beam enters into a dispersive element (the prism or grating). In order to scan the necessary spectral range the dispersive element is turned in some angle. The light beam passed through the prism is focused in the plane of the output slit. The input and output slits of the monochromators are placed vertically. To compensate the change of light energy at the change of wavelength, the slit width is usually changed during scanning. This technique is unacceptable in the case of Fourier spectrum recording because the beamwidth is changed. In order to compensate the energy change of the output beam the ordinary spectroscopic cell with dyes can be used as an attenuator. The input slit restricts the angle in the horizontal plane. Light is propagated within the limits of this angle. The decrease in the slit width leads to the decrease in the spectral width of the radiation. Furthermore, the image of the input slit should not exceed the sizes of the recording photodetectors. This defines the maximal established adjusted slit width.

It should be noted that some degree of polarization of radiation is typical for monochromators and should be taken into account. When we operate with the guided modes of certain polarization while performing measurements. For this reason a polarizer was inserted into the device, which allows one to get a light beam with a specified direction of polarization (specified plane of electromagnetic field electric component oscillation). The light beam enters into the prism coupler after passing through the polarizer and the lens. The investigated structure is positioned in the focal plane of the focusing element. The sample under test is set in contact with the measuring prism in order to provide the optimal conditions for the excitation of the optical mode in the sample. The radiation reflected from the sample passes through the polarizer or the filter, the focusing element performing Fourier transforms and is then recorded by the photodetector. The measurement of the spatial distribution of the intensity of the reflected light beam is performed with the help of a CCD-array photodetector (12), the rotation axis being conjugated with the axis of the rotary table. The recording plane of the photodetector is placed in the focal plane of the lens (13), through which the reflected light beam passes. The power of the incident light beam is controlled with the help of the photodetector (11). Sluggishness of the sample response can be measured with the help of different modulators using stationary light sources or time structure of the pulse light sources. Note that the strict positioning of the investigated sample in the focal plane of the optical element (13) is not required. The measurement results would not change in case the base of the prism coupler is positioned in the focal waist of the beam. That is, the chromatic aberration of the lens is not strict.

While recording the spectra, we should take into account the hardware-controlled function of the equipment including the sensitivity of the recording device to different wavelengths. It is very difficult to calculate this function. As such this function is measured with the help of sources with known absorption spectrum or by using calibrated photodetectors with known spectral sensitivity. A lamp with a special tungsten tape heated to some temperature, whose radiation coincides with the spectrum of a black body with an accuracy of some constant, can serve as such a source in the visible spectral range. Therefore, it is possible to find the hardware-controlled function of the device by using the measured and calculated radiation spectra of a tungsten lamp. This calibration is quite a complicated and it cannot be solved if one tries to get the information of the qualitative behavior from the spectrum (at the level of the presence or absence of absorption bands).

While investigating the absorption spectra, all the complicated phenomena taking place in thin-film structures become apparent. These phenomena make the interpretation of the obtained results difficult. One of the typical features of many semiconductor films is the nonlinearity of their optical properties, which takes place even at very small intensities of the exciting light. The processes in such structures depend on the photon number of the exciting light that are absorbed in the volume unit of the film. If one does not take special measures, the intensity of the exciting light is greatly changed by varying the radiation wavelength. Since the excitation intensity is changed at the excitation by quanta of different energy the separation of the role of the quantum energy from the general picture becomes a complicated problem. For this reason the device is supplied with an equipment that stabilizes the intensity of the incident light at different wavelengths by controlling the filament current of light source. But the problem of creation of equal power density of the exciting light absorbed by the volume unit is solved partially. This situation takes place because the density and distribution of the absorbed energy are changed with the variation of the absorption coefficient at different wavelengths. Thus, the error in the determination of the absorption coefficient increases with the variation of the wavelength of the incident light. In this case it is possible to measure the spectra at the constant photon number in the incident light beam. Measurements of such type require a high degree of device automation because the amount of the processed data is very large. As a result the signal enters into the on-line storage after the digital processing. The instrumental error of determination of the mode excitation angle is equal to 2×10^{-5} rad, and of light beam intensity is about 0.1%.

Devices of such type allow one to investigate various characteristics of thin films in non-coherent light.

6.3. Special Features of Absorption Spectra Recording by Waveguide Spectroscopy and their Processing

The peculiarities involved in the observation and processing of the recorded signals are similar to the case considered in Chapter 4. Let us now consider the fact that the light beam passed through the exit slit does not always have Gaussian intensity distribution over spatial frequencies in the direction normal to the slip. This fact is often caused by the excessive slit opening or poor adjustment of the device. It therefore useful to measure the hardware-controlled function of the device. Taking into account all these peculiarities one can approximate the output light beam by the Gaussian distribution.

There is a significant issue involved here. The forming of the spatial distribution of the beam intensity is concerned with the interference of the light beam reflected from the prism coupler and the beam passing along the waveguiding film and reflected back into the prism. In order to determine the conditions of the recording of the absorption spectra in quasi-coherent light the influence of the coherence on the accuracy of the determination of h'' were investigated performed (Figure 6.2).

From expression (3.1.11) it follows that the coherence length for radiation of monochromator with $\Delta\lambda = 3$ nm is equal to ~ 130 µm, on the other hand the length of the mode track propagating along the waveguide at $h'' = 3.1 \times 10^{-4}$ is equal to ~ 150 µm. The light propagated along the film is radiated into the prism and takes part in the creation of the interference

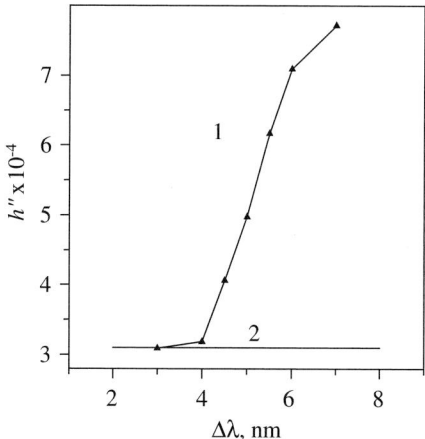

Fig. 6.2. Dependence of h'' on the radiation spectrum width (1); line (2) is the measured value of h'' with the help of single-mode He–Ne laser with wavelength of 0.633 µm.

picture. As we can see these parameters have very similar values, this allows one to use the quasi-monochromatic radiation for the determination of thin-film parameters.

Here, it should be noted that the stated method could be used for determining parameters of the probe light beam such as coherence length.

The second peculiarity of these measurements is the fact that depending on the real range of the recorded frequencies the signal spectrum is always "fuzzy" to some extent. This means that the signal spectrum contains the lines, which differ by the amplitude and frequency from those that really exist in the signal [187]. Because of this there is a problem of formulation of the reliable ideas of the properties and characteristics of the analyzed object on the basis of the spectrum obtained in the limited range of spatial frequencies. It should be mentioned that the obtained results can be satisfactorily correlated with the data obtained from the application of well-known methods. Therefore, if we use model approximation of the process of the light beam reflection when the guided mode is in the excited states, as described in this chapter, we can overcome the effects caused by the limited range of data recorded in the experiment. Usually, this leads to masking of the weak effects and to the restriction of the resolution and sensitivity [188]. In the application of the gradient descent technique for the determination of guided mode propagation constants (see Chapter 3), we can determine the h values using the information about the whole array of experimental points of the angular Fourier spectrum. Unlike the method described in Chapter 4, which uses only the values of extremuma in the intensity distribution, i.e. values of I_{max}, I_{min}, φ_{min} and angular width of m-lines, in this method the finiteness of the range of angles φ, where the intensity is recorded, is significant. This brings some errors in the determined values of h but allows one to process the non-symmetrical distributions of $b(\varphi)$, which is very useful in our case. Preliminary measurements of thin-film parameters showed that the angular shift of the beam center within the limits 30–40% of its width leads to a measurement error of h'' in the limits of 3.5%. This is quite a satisfactory value for films with the absorption of $50\,cm^{-1}$ and thickness $\sim 0.5\,\mu m$.

6.4. Measurement of Absorption Spectra by the Fourier Spectroscopy of Guided Modes

The first attempts to measure the spectral absorption coefficient of thin films were made in the spectroscopy of surface electromagnetic waves [190], which allowed one to investigate the vibration spectra of monomolecular layers and oxide layers on metal surfaces. This method is quite effective in the infrared spectrum region. The method proposed the use of the modes of the optical waveguide in the visible range in absorbing film spectroscopy [191]. While performing measurements by this method the films were placed on the surface of the waveguide and the additional light attenuation in observed the waveguide was caused by the light absorption in the investigated film as a result of penetration of the field of the guided mode into the film. This method gave satisfactory results for strongly absorbing extra thin films (10–15 Å). In order to determine the spectral absorption coefficient of the thin film, the real h' and imaginary h'' parts of the propagation constant h of the guided mode were determined with the help of a two-prism scheme [192]. The necessity of small losses in measurement makes one to resort to various tricks. For example, the thickness and refractive index are measured by the waveguide methods, and the absorption coefficient is determined by the film transition photometry taking into account the data obtained during the waveguide measurements [193]. Unfortunately, the necessity of the additional adjustment of the prism coupler after reconfiguration of the incident radiation over the frequency and the dependence of h'' on the coupling power have restricted the application of these methods.

Some of these problems can be removed by the application of Waveguide Spectroscopy technique stated in this chapter.

The analysis of the experimental results shows that if we use the appropriate optical scheme and reduce the spectral bandwidth of radiation to the value of <5 nm, then the error in determining the mode propagation constant relative to the value of h'', measured at the wavelength of the laser radiation, does not exceed 1%. It is obvious that the influence of the light coherence on the accuracy in determining the parameters of the guided mode, requires additional investigations because the formation of the intensity distribution, in the cross-section of the reflected light beam is concerned with the interference of light. But the measurement results show that this method allows one to determine correctly the spectral absorption coefficient of thin films in a quite wide wavelength range [194]. The reconfiguration of the frequency of the probe radiation causes the transformation of the spatial distribution of the intensity of the reflected light and the recorded dependence becomes unsymmetrical. The application of steepest descent technique allows one to determine the real h' and imaginary h'' parts of the propagation constant in case of unsymmetrical intensity distribution of the reflected beam. It allows one to find the refractive index, absorption coefficient and film thickness using the values of h for different modes. It is worth noting that the method used for determining h mentioned Chapter 4, in this case takes into account the influence of the prism coupler and gives the possibility to determine those parameter values that do not depend on the conditions of the experiment. This method does not require to measure the absolute magnitude of the light intensity. The application of the wide (in the wavevector space) light beam and the use of steepest descent technique while determining h allows one to avoid additional adjustment of the prism coupler while reconfiguring the incident radiation over frequency. As a result, one can determine the film absorption coefficient, refractive index and its thickness in the specified spectral range.

Let us consider the results of the investigation of parameters of films obtained by RF sputtering of ceramic target made from zinc oxide on quartz glass substrate at $T = 300$ K. This example allows one to illustrate the abilities of the waveguide spectroscopy of thin films. The thickness of the film is equal to 1.5 µm and can also the propagation or five guided modes. The $n(\lambda)$ and $k(\lambda)$ dependencies for the thin film are depicted in Figure 6.3. These dependencies are determined by the waveguide method, where k is related to the absorption coefficient α (cm^{-1}) by the expression $\alpha = 4\pi k/\lambda$. The method allows one to record the absorption coefficient about 1 cm^{-1} ($k = 2 \times 10^{-6}$) in films with thickness <1 µm. For quite thick semiconductor films ($d \sim 1$ µm) the measurements become easier because in this case $k \approx h''$ [71], and as it is evident from the analysis of the curves given in Figure 6.3, the spectral absorption coefficient is characterized by the dependence $h''(\lambda)$.

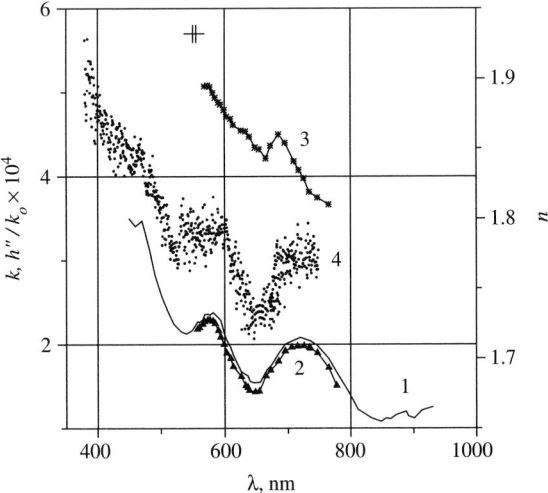

Fig. 6.3. Spectral dependencies of h'' (1), the absorption coefficient k (2) and the refractive index n (3) for the film ZnO; the $k(\lambda)$ dependence measured by traditional spectroscopy technique (4).

That is, in order to determine the film absorption coefficient it is enough to measure the values of $h''(\lambda)$. The requirement of the existence of guided modes in the tested structure is not compulsory when the film is deposited on the weakly absorbing substrate, i.e., when the absorption coefficient of the substrate material is much less than that of the films. The substrate refractive index can exceed the film index in such a structure. It is possible to determine the film parameters by measuring the propagation constants of leaky modes. There are problems in determining $k(\lambda)$ values less than 10^{-5} while using the waveguide spectroscopy method because of the restrictions of this method concerned with the "leakage" of the guided mode energy from the prism coupler. This has already been mentioned in the previous chapters and it should be taken into account while performing the measurement of the absorption coefficient of weakly absorbing films.

The measurement of spectra in the visible range allows one to investigate the influence of the doping level on the film's properties. The dependencies of $h''(\lambda)$ for $SnO_2 : Sb_2O_5$ films with different proportions of antimony are given in Figure 6.4.

The dependencies for films $SnO_2 : Sb_2O_5$, $SnO_2 : Al_2O_3$ and $SnO_2 : WO_3$ (the fundamental mode is of TE polarization) are depicted in Figure 6.5. Measurements were performed for films $SnO_2 : Sb_2O_5$, $SnO_2 : Al_2O_3$ and $SnO_2 : WO_3$ deposited on the quartz glass substrate at temperature 420 K. The $SnO_2 : Sb_2O_5$ film thickness is equal to 0.7 μm and two modes of TE polarization propagate in the film. From the analysis of the spectral

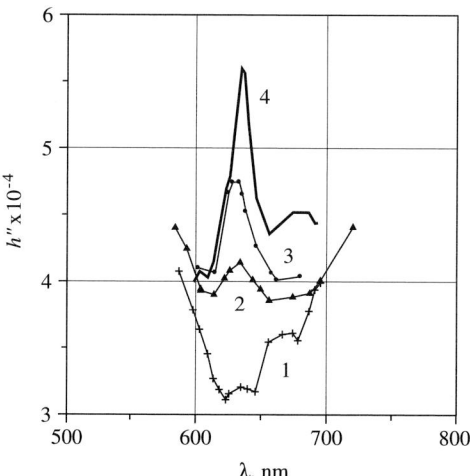

Fig. 6.4. Dependencies of $h''(\lambda)$ for SnO_2 films with different concentrations of antimony oxide $C_1 < C_2 < C_3$, (curves 1–3, respectively), in air (3) and air medium with ammonia impurity (4).

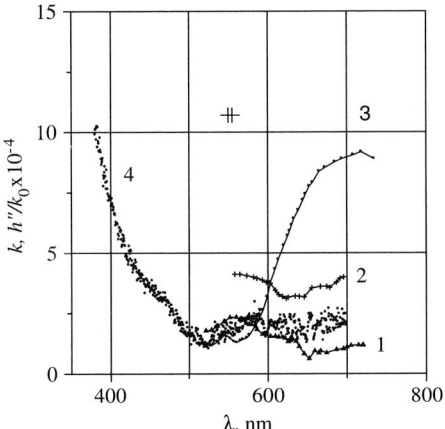

Fig. 6.5. Dependencies of $h''(\lambda)$ for SnO_2 films with different type of doping impurity SnO_2: Sb_2O_5 (1), SnO_2: Al_2O_3 (2), SnO_2: WO_3 (3); spectral dependence of absorption coefficient for SnO_2:WO_3 (4) film.

dependencies it is evident that the change in the impurity material and the change in the doping level displayed themselves in the structure and the position of absorption band, which are recorded in transmission range of these films.

The sensitivity of this method can be demonstrated by the following example. It is known that the films of tin dioxide are widely used as an active element in gas sensors [195]. The electrical properties of the film are changed during gas adsorption in the film with the thickness of one molecular layer of the gas molecules. It is natural to assume that the optical parameters of the film will undergo some changes. Spectral dependencies of $h''(\lambda)$ for the $SnO_2 : Sb_2O_5$ film of definite compound in the air atmosphere and air mixture with ammonia impurity (curves 3 and 4, respectively) are depicted in Figure 6.4. The presence of ammonia impurity with the concentration of about 0.01 mg/L causes changes in film parameters. This fact is recorded in film absorption spectrum. The total relative error in determining $k(\lambda)$ taking into account the systematic errors does not exceed 0.03. Spectra of the ZnO film measured twice with the help of traditional methods of spectrometry under identical conditions are depicted in Figure 6.6.

Two spectra of the same film determined with the help of waveguide spectroscopy are depicted in the same figure. Analysis of the depicted dependencies shows good repeatability of results of this method [196].

In order to check the feasibility of the results obtained by the above method the measurements of dye "Methyl Red" absorption spectra were

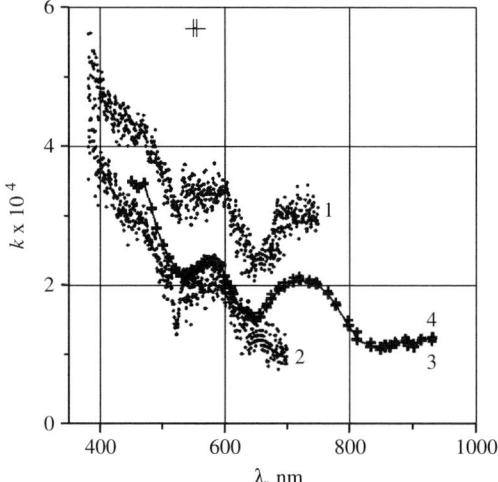

Fig. 6.6. Repeatability of ZnO film absorption coefficient measurement results obtained by traditional spectroscopy method (1, 2) and waveguide spectroscopy (3, 4).

performed. This dye is traditionally used in optics and its properties are well known. It is known that the absorption spectrum of this dye is stable and the position of the lines practically does not depend on the host [197]. The measurements of absorption spectrum of acrylic resin film doped by dye show a satisfactory agreement of the results with the data obtained by well-known methods (Figure 6.7).

In addition, the measurement of the spectral absorption coefficient was performed for the investigated oxide films. The values of $k(\lambda)$ for zinc oxide film are depicted in Figure 6.4 (curve 4) and for film $SnO_2:WO_3$ in Figure 6.5 (curve 4). A good correlation of the shape of spectral dependencies obtained by both the methods is observed for the given curves. The maximum discrepancy of curves $k(\lambda)$ and $h''(\lambda)$ for thin films is recorded in the long-wave spectral range (Figure 6.5). This can be explained from the point of view of the waveguided mode theory [9], since there is a decrease in the effective thickness d/λ of the film with increase in radiation wavelength. In our case the $SnO_2:WO_3$ film thickness is equal to 0.3 μm.

The determination of film thickness during waveguide measurements allows one to estimate the accuracy of the obtained results in comparison with the obtained value of the film thickness measured by an independent method, for example, by multibeam interferometry or with the help of mechanical stylus (Figure 6.8). The value of the film thickness $d = 1.501 \pm 0.01$ μm; the thickness measured with the help of the mechanical stylus was equal to 1.50 ± 0.02 μm.

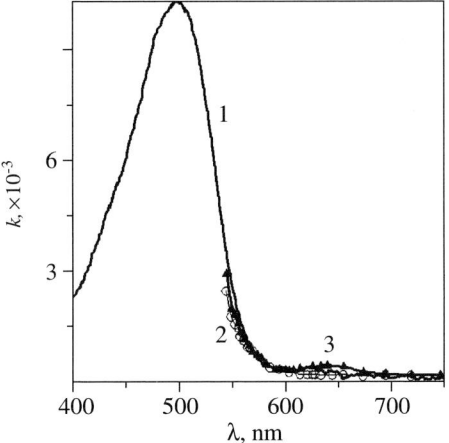

Fig. 6.7. Absorption spectra of polymeric film doped by dye Methyl Red measured by photometry technique (curve 1 according to Dr A.I.Voitenkov), waveguide spectroscopy for TE and TM waves (curve 2 and 3, respectively).

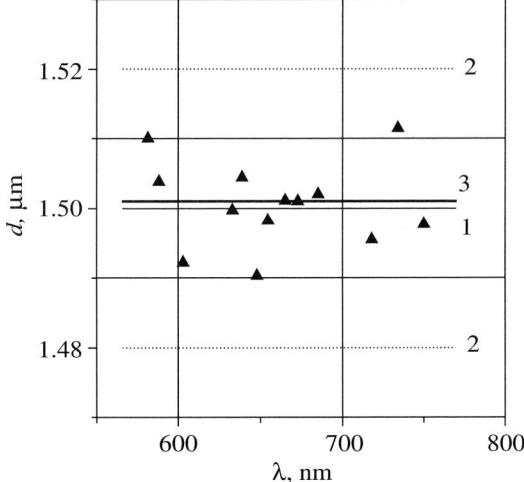

Fig. 6.8. The ZnO film thickness measured with the help of mechanical stylus (1), by waveguide method (3) and error of thickness measurement by mechanical stylus (2).

Taking into account all the stated results we can now discuss on the creation of the basis of a new method that can be used for the investigation of optical parameters of a thin film – the Waveguide Spectroscopy.

Chapter 7
Applications of the Waveguide Spectroscopy Techniques in Sensor Systems

7.1. Integrated-Optics Sensors and their Features 128
7.2. Gas Thin-Film Sensors 134
7.3. Physical Origin of Processes on Surfaces of Thin-Film Sensors 139
7.4. Evaluation of the Adlayer Parameters................................. 146
7.5. Sensors with Recording of the Light Beam Reflection Coefficient 147
    7.5.1. Gas Sensors 147
    7.5.2. Integrated-Optics Sensors of the Angular Movement 149
7.6. Waveguide Microscopy of Thin Films.................................. 151

The development of the techniques for studying the optical properties of thin films and then appropriate equipments considered in the previous chapters form the basis of waveguide spectroscopy of thin films. The name "*waveguide spectroscopy*" specifies the way of gaining information about the thin-film properties, when the optical modes are excited in the studied thin-film structure. The mode characteristics are measured and the optical parameters of thin films are determined with the help of experimental data.

The high accuracy of the measurement methods of the film parameters allows one to consider the structures based on a prism coupler as the potentially sensitive waveguide device of different functionality. This, in its turn, opens up the possibilities to create sensors of new type. The enhancement of the existing methods of inspection and the analysis of the environment, and the development of new methods are important problems common to all fields of human activity. In recent times, hundreds of sensors of different types and applications have been designed and manufactured. In addition to having high metrological characteristics the sensors must have high reliability, long operating life and stability, small size, mass, and low-power consumption. Besides, the sensors must be compatible with microelectronic devices of data processing and should have low manufacturing cost. Integrated-optics gas or gas impurity sensors satisfy these requirements to some extent. Optical sensors are also quite perspective devices for application in the systems of remote monitoring of environment parameters, especially under conditions of high radiation hazard, because they give the possibility to connect the sensitive element with the recording system by optical fiber [198]. The examples of the applications of thin-film structure based on a prism coupler as the optical sensor for the measurement of the environment parameter are given below.

7.1. Integrated-Optics Sensors and their Features

The real part h' of the mode propagation constant is one of the basic physical characteristics of the guided mode in thin-film waveguides, which are usually the sensitive elements of integrated-optics sensors. As we already know, the propagation constant depends on the radiation polarization (TE or TM), mode number, wavelength and such parameters of a guiding film as the refractive index n, reduced thickness (i.e. ratio of the film's thickness to the light wavelength), the substrate and surrounding refractive index n_s and n_c, respectively. The physics of the processes taking place in integrated-optics sensors is concerned with the interaction of the electromagnetic field of the optical guided mode with the surroundings. At each internal reflection of the light in the waveguide the interference between the incident and the reflected beams creates the non-propagating standing wave, which is normal to the reflecting surface. The energy associated with this evanescent wave tails out into the surroundings, where it can interact with gas molecules. The depth Δy of the field penetration into the surrounding, which covers the waveguide and can serve as operating layer of sensors, is defined by the following expression [199]:

$$\Delta y \equiv (\lambda/2\pi)[(h'/k_0)^2 - n_c^2]^{-1/2}. \quad (7.1.1)$$

Here, it should be noted that the decrease in the amplitude of the penetrating field with the gradual increase in the distance from the waveguide surface is described more correctly by the exponential law. The field of this wave "feels" changes in the refractive index near the waveguide surface. This appears in changes in the mode propagation constant [199]. Two different effects can cause these changes:

- The formation of the adsorbed or bounded molecule layer. The molecules are carried from the volume of the gas or liquid surroundings to the waveguide surface with the help of convection or diffusion. This adsorbed layer is modeled as the homogeneous layer, which has thickness d_1 and refractive index n_1.
- Some changes in the refractive index Δn_c of the homogeneous (liquid) sample covering the waveguide surface.

In the case of the microporous waveguiding film there is one more effect: adsorption or desorption of molecules by the pores of the waveguiding film. These processes change the film refractive index n and this index variation Δn causes the change in the magnitude of the mode propagation constant $\Delta h'$. The distribution of the mode field inside the waveguide is responsible for this effect. If all the factors listed above simultaneously affect the guiding structure, then the resulting change in the real part of the mode propagation constant is defined by the following expression:

$$\Delta h' = \left(\frac{\partial h'}{\partial d_1}\right)\Delta d_1 + \left(\frac{\partial h'}{\partial n_c}\right)\Delta n_c + \left(\frac{\partial h'}{\partial n}\right)\Delta n. \quad (7.1.2)$$

Derivatives $(\partial h'/\partial d_1)$, $(\partial h'/\partial n_c)$ and $(\partial h'/\partial n)$ depend on the optical parameters of the waveguide, substrate, and the sample under test. The first of the effects underlying the operation of the integrated-optics sensors makes it possible to observe the adsorption or desorption of gas molecules at the surface of waveguides in the real time. The variation of the refractive index is the basis of application of the sensors as different refractometers. Both these effects are the basis for the development of the bio- and chemical sensors. The filled micropores can be used for measuring relative humidity and for determining the presence of gas impurities [200].

The practical realization of sensing elements used for the determination of physical parameters of the media is based on the use of the planar optical waveguides and application of the gratings or prism couplers. The basic characteristics of any sensor are its sensitivity, resolution, stability and validity of the obtained results. The value S is the sensor sensitivity that characterizes the changes in the measured parameter when there is variation in the concentration of recorded impurity C in the environment. Most often, the real part h' of the mode propagation constant is the measured value in integrated-optics sensors; thus

$$S = \Delta h'/\Delta C \quad \text{or} \quad S = dh'/dC. \quad (7.1.3)$$

It is obvious from this expression that

$$h' = SC + b, \quad (7.1.4)$$

where b is the propagation constant in the absence of impurity in the environment.

Expression (7.1.4) is the equation of line (Figure 7.1), which is referred to as the calibration line. Coefficients S and b can be determined using the mathematical techniques of approximation and optimization [164, 189]. For example, if N measurements are performed at different values of concentration C_i, then, using the least-squares method [201], one can obtain

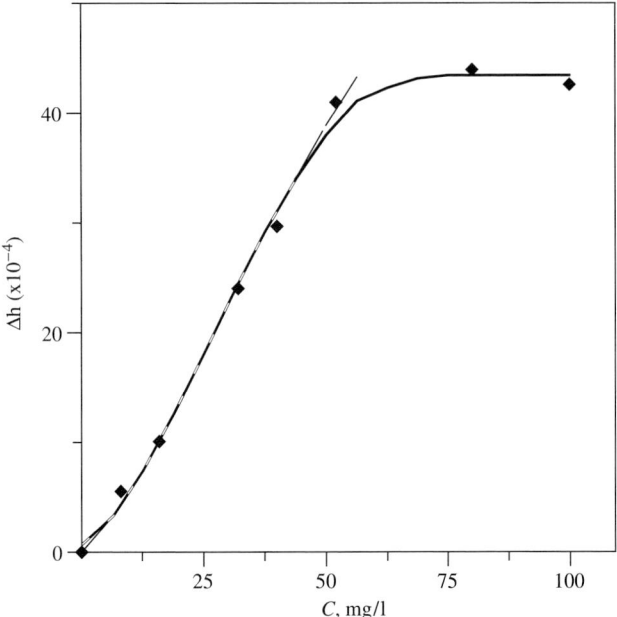

Fig. 7.1. Dependence of the measured parameters $\Delta h = h' - b$ on the concentration C.

$$S = \frac{N\sum_i C_i h'_i - \sum_i C_i \sum_i h'_i}{N\sum_i C_i^2 - \left(\sum C_i\right)^2},$$

$$b = \frac{\sum h_i - b\sum C_i}{N}.$$

Stability is a parameter describing the random errors of measurements and characterize the dispersion of repeated measurements. Deviation from the result obtained during the next measurement from the mean value $\langle h' \rangle$ for the series of measurements h_i, or the difference between the maximum and minimum of measured values can be used as the criterion of stability. Usually, in experiments, the dispersion of measured values follows the Gaussian (normal) distribution. More detailed information about the special techniques used for the determination of the distribution type of the results of measurement of the analog signal can be found in Refs. [165, 202–204]. For the estimation of stability the dispersion σ^2, root-mean-square deviation (standard deviation) σ and relative deviation δ_R should be calculated. For a large measurement series (theoretically $N \to \infty$), σ determines the absolute measurement error δ. In practice, when the number of possible measurements is always limited, the errors of the single measurement do not exceed the doubled root-mean-square deviation in most cases (see Section 3.2).

The minimum concentration $C_{\min}$, which can be detected by the device with some confidence probability, is considered as its resolution:

$$C_{\min} = \frac{\Delta h'_{\min}}{S}, \qquad (7.1.5)$$

where $\Delta h'_{\min}$ is the lower limit of the measured value of changes in h'.

To begin the measurements it is useful to estimate $\Delta h_{\min}$ with the help of the statistical criterion

$$\Delta h_{\min} = \bar{b} + K\sigma, \qquad (7.1.6)$$

where $\bar{b}$ is the mean value of h' without impurities in the environment and σ is its standard deviation.

The coefficient K characterizes the confidence probability, and is usually equal to 2 or 3.

Thus, the measurement resolution can be written as

$$C_{\min} = \frac{K\sigma}{S}. \qquad (7.1.7)$$

The impurity concentration values determined in practice greatly exceed the lower limit. As a result, the important parameter of the sensing device is the dynamic range, i.e. the range of concentration values, which can be determined by the given technique. The minimum value of this range is conditioned by the resolution, and the maximum value of this range is caused in most cases by the linear range of the calibration curve (see Figure 7.1).

While performing measurements there is the problem in the feasibility of the obtained results, which is the parameter characterizing the closeness of the obtained and real values of the measured quantity. The validity of measurements is usually characterized by the systematic error caused by the parameters of the device used in experiments and errors of the method. The application of the inadequate models, approximations during the calculation, etc. can be the origin of the systematic error. We can mark the following techniques of determining systematic errors: the use of standard sample, the method of additives and the method of comparison of measurement results obtained by independent techniques [205].

Further, we will consider the examples of the existing integrated-optics sensors. The grating in sensor systems can be used for the input of the light into the waveguide (Figure 7.2), and also for the output. The angle α in the given scheme is the optimal angle of the guided mode excitation. The

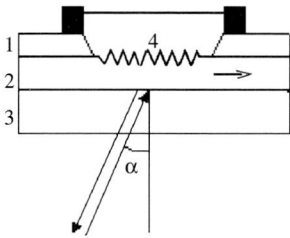

Fig. 7.2. Grating sensor structure: dielectric layer (1), waveguiding film (2), substrate (3), sample under test (4) [199].

radiation is brought out from the waveguide in the scheme under the same angle with the help of the grating.

The relationship between the parameters of the guided mode and the parameters of the used grating is given by the following expression:

$$N_{\text{ef}} = n_a \sin \alpha_l + l\lambda/\Lambda, \qquad (7.1.8)$$

where $N_{\text{ef}} = h'/k_0$ is the effective refractive index of the guided mode, λ the wavelength and Λ the grating period and $l = \pm 1, \pm 2, \ldots$ the diffraction order [206].

As shown earlier (see expression (7.1.4)) the changes in the guided mode propagation constant are caused by the variations of the properties of the medium. Thus from (7.1.8), the following expression can be obtained:

$$\Delta N = n_a \Delta(\sin \alpha_l) \approx n_a \cos \alpha_l \Delta \alpha_l. \qquad (7.1.9)$$

A change in N_{ef} leads to a change in $\Delta \alpha$ of the angle α_l. As ΔN is the value characterizing the changes in the propagation constant in the area of grating, the parts of the waveguide lying outside the grating can be covered with the dielectric layer. Such sensing structure was used as a biosensor for the real-time observation of the adsorption of the protein on the waveguide surface. The possibility of its application as the sensor of relative humidity was also studied; the instrumental resolution in the used scheme was $\Delta N_{\min} = 2 \times 10^{-6}$ [207–211]. Similar results were obtained when the grating was used as the light output device [212].

There are different types of sensors, whose operation is based on the interferometer principle. They can be conditionally classified as the single- and double-beam interferometers. The realization of the first type is described in Refs. [213–215], and sensors based on the double-beam interferometer (Figure 7.3) are considered in Refs. [216–219].

The laser radiation ($\lambda = 632.8$ nm) is splinted into two similar beams, which enter into the sensor and into the comparative channel of the waveguiding structure. The change in parameters of the sensor channel caused by

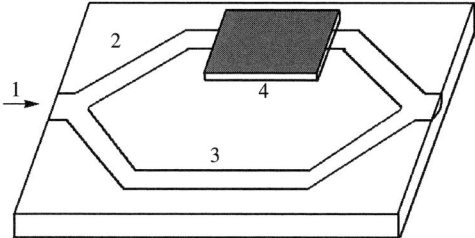

Fig. 7.3. Gas sensor based on the Mach–Zehnder interferometer: input light (1), sensor and comparative chanel (2, 3), sensitive layer (4).

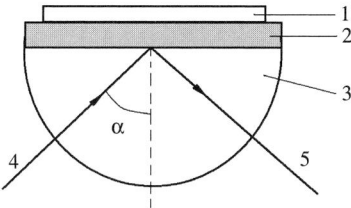

Fig. 7.4. SPW sensor: gold film (1), gap (2), prism coupler (3), incident light beam (4), reflected light (5).

the adsorption of the investigated substance leads to the difference in the optical lengths of the channels because the properties of the comparative channel are not changed. The phase shift is written as

$$\Delta\phi = L\frac{2\pi}{\lambda}\left(\frac{\partial N_{\text{eff}}}{\partial d_{\text{f}}}\Delta d_{\text{f}} + \frac{\partial N_{\text{ef}}}{\partial n_{\text{c}}}\Delta n_{\text{c}}\right), \qquad (7.1.10)$$

where L is the length of interaction, n_{c} the refractive index of the cover, and d_{f} the adlayer thickness.

The measurement limit obtained experimentally was $\Delta \tilde{N}_{\min} = 2 \times 10^{-8}$ [219]. Some advantages of these sensors are high sensitivity and processing stability.

The operating principle of the surface plasmon wave (SPW) sensor is similar to the functioning of integrated-optics sensors: the sensing effect appears here as a result of the SPW interaction with the investigated substance (Figure 7.4). Because of the analogy in the way of the excitation of the guided modes and SPW by the prism and grating couplers it is possible to use a similar structural layout while creating a sensor.

The optical excitation of the surface plasmons by the method of attenuated total reflection was demonstrated by Otto [97] and Kretschmann [98]. Particularly, the Kretschmann geometry of the method has been found to be suitable for sensing elements and has become the most widely used geometry in SPW sensors. In this configuration, the light wave is totally reflected at the interface between the prism coupler and the thin metal layer and excites the SPW at the outer boundary of the metal by tunneling through the thin layer of the metal. In this case the prism coupler can be isolated from the investigated substance. The biosensing structure described in Ref. [220] can serve as an example of the technique described above. Many gas detectors are based on the effect of SPW excitation [220–226]. The comparison of sensitivity of the SPW and the integrated-optics sensors is given in Ref. [221]. The grating couplers were used in both schemes for the excitation of electromagnetic waves. The sensitivity of the SPW sensor while determining the changes in the refractive index was an order worse than that of the integrated-optics sensors ($\Delta N_{min} = 2 \times 10^{-4}$ for the SPW sensors and 3×10^{-6} for the integrated-optics sensors).

Nowadays, a number of sensors are used for the determination of the physical parameters of the medium, which are fabricated on the basis of planar optical waveguides to be used as the grating of the prism couplers [227, 228]. The operating principle of these devices is based, as usual, on the recording of the resonant angles of the mode excitation. But the changes in the angles, caused by an external action on the waveguide, are insignificant. This makes it difficult to process the measurement results. Besides, the resonant angle of the guided mode excitation is concerned with the real part of the mode propagation constant only and gives no information about its imaginary part, which is also changed due to external action.

7.2. Gas Thin-Film Sensors

The gas sensors considered in this chapter are based on a prism coupler, the recorded quantity [229, 230] being the integral intensity or the spatial distribution of the intensity of the light beam reflected from the prism coupler at the specified incidence angle [231, 232]. The prism coupler is the resonator that differs from the classical Fabri–Perrot resonator. The first difference is the simplicity of the resonance excitation realized by changing the angle between the light beam axis and the normal to the prism face, and the other difference is the wide range of the variation in the quality factor provided by a mechanical shift of the prism in the direction normal to the investigated surface. The estimations show that due to the resonant properties of the considered structure the noticeable changes in the spatial distribution of the reflected light beam intensity take place at the small variation of the complex refractive index of the film, e.g. of the value $\sim 10^{-6}$. Such changes appear in the variation of the composition of the surroundings.

Thus, it is natural to make an attempt to determine the parameters of the medium and the layers, adsorbed on the surface of the waveguiding structure by using the measurement methods described above. The reflected radiation intensity and its spatial distribution depend on the variations of the real and the imaginary parts of the mode propagation constant [170]. Devices based on the prism coupler with the waveguiding structure deposited on its base are characterized by the significant amplification of external actions. Besides, these are rigid structures with a long-term operation reliability. The sensing element can be fabricated as the thin film deposited on the base of the prism coupler. The principal scheme of such sensor is depicted in Figure 7.5.

The prism coupler in the form of the isosceles glass prism (1) is the substrate and is also used for the excitation of the guided mode in the semiconductor film (3). The film is separated from the prism by the dielectric gap (2) (the silicon dioxide film). The dielectric gap is used for providing the waveguiding regime in the conductor film and also for the optimization of the sensor parameters. The sensing layer (3) is obtained by sputtering of the ceramic target fabricated from the mixture of tin oxide and antimony oxide. The concentration of antimony oxide ranged from 2 to 15 wt% [233]. These films were deposited by RF sputtering in the argon and oxygen gas mixture. The thickness of the dielectric gap obtained by RF sputtering of the quartz glass target varied in the range from 0.2 to 1.0 µm, and that of sensing layer from 0.07 to 1.5 µm. The light from the radiation source ($\lambda = 0.633$ µm) is coupled into the conductor film and excites the guided mode. The complex propagation constant h of the guided mode depends on the optical and geometrical parameters of the gap, the waveguiding film and environment properties. The appearance of the detected gas impurity (4) in the environ-

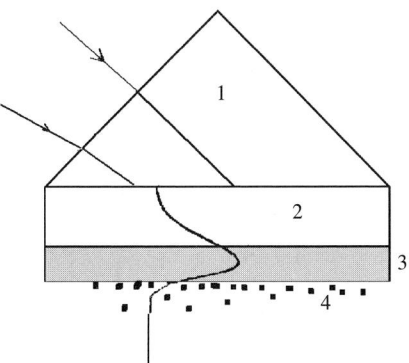

Fig. 7.5. Prism-based sensor.

ment leads to the variation of the optical parameters (the refractive index and the absorption coefficient) of the waveguiding film. This causes the changes in the mode propagation constant, which appeared in the modification of the Fourier spectrum of the reflected light beam. The optoelectronic circuit (see Section 3.4) consisting of the photodetector array and the analog–digital converter records the spatial distribution of the reflected light intensity and allows one to measure the changes [160]. The real h' and the imaginary h'' parts of the mode propagation constant are determined during the processing of the intensity distribution. The changes in these values can serve as the measuring parameters of sensor sensitivity.

Let us consider the results of studying the described structure depending on the contents of ammonia, alcohol and acetone vapors in the environment. In all such cases the temperature of 20 °C and relative humidity of 80% were constant in the analyzed volume. The concentration of the impurity in this volume can be measured with the help of a flowmeter. The vapor concentration in the analyzed volume can be determined if the liquid mass m in the surrounding and the volatility of the given substance are known. The volatility L (mg/L) is determined from the expression

$$L = 16pM/(273 + t),$$

where p is the pressure of the saturated vapor and M and t are the molecular mass of the substance and the temperature of the environment (°C), respectively.

The saturated vapor pressure can be found in the literature. If p is not known then it can be determined from the expression [240]

$$\lg p = 2.763 - 0.019/t_b + 0.024t,$$

where p is the saturated vapor pressure (in mm of mercury) and t_b the boiling point (°C).

The results of the study of the sensor sensitivity to the impurities in the gas medium are depicted in Figures 7.6 (a) and (b), where $\Delta h''/h_0'' = (h'' - h_0'')/h_0''$ and h'', h_0'' are the imaginary parts of the mode propagation constant in the presence of impurity and in the air, respectively. If the real part h' of the mode propagation constant changes insignificantly within the range of concentration variations, then the values of $\Delta h''$ allow one to use h' as the sensitivity parameter and such prism-based structure as the gas sensor. The proposed structure has the highest sensitivity to ammonia impurity, the recorded concentration being of two or three orders less in comparison to the concentration of other gases, i.e. a sensor of such type has quite high selectivity. The range of recorded concentrations is 10^{-4}–10^{-6} vol%. The sensor sensitivity and the range of the recorded concentrations can be changed by varying the physical parameters of the thin-film structure. For

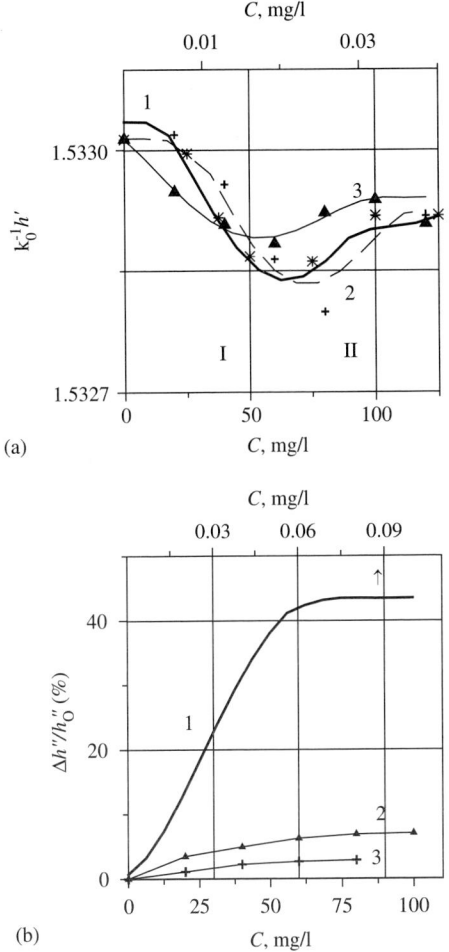

Fig. 7.6. Dependence of real (a) and imaginary (b) parts of the mode propagation constant on concentration of detected gas for ammonia (1), ethyl alcohol (2) and acetone (3).

example, the decrease in waveguide losses of the semiconductor film, i.e. of the h_0'' value, leads to the increase in the sensor sensitivity (Figure 7.7). It is obvious that the range of the recorded concentrations decreases in this case. Similar results can be obtained if the gap parameters are changed [230]. The questions related to the optimization of the construction of the sensitive element are considered in details in Refs. [230, 234, 235].

In order to evaluate the sensitivity of such a structure, one can consider the results of measuring electroconductivity of the same films, performed by

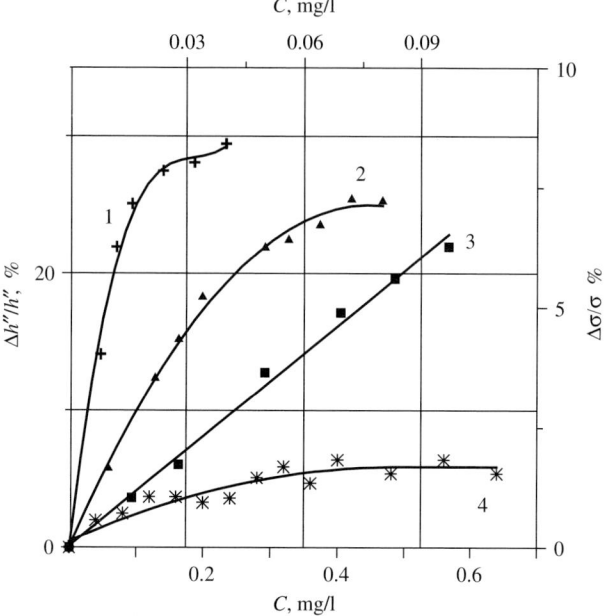

Fig. 7.7. Changes in the sensor sensitivity with the decrease in the optical losses in the waveguiding films: $h_0'' = 1.82 \times 10^{-4}$ (1), $h_0'' = 3.2 \times 10^{-4}$ (2), $h_0'' = 4.87 \times 10^{-4}$ (3); and in conductivity (4) for the structure in ammonia environment.

the electrical four-probe technique. The dependence of the conductivity variation $\Delta\sigma/\sigma_0 = (\sigma - \sigma_0)/\sigma_0$ on the concentration of ammonia in the air, where σ and σ_0 are the film conductivities in the presence and absence of impurity, respectively. These measurements were also performed at room temperature. Owing to the lower sensitivity of the electrical parameters to the variations of the composition of the medium surrounding the sensor, the measured concentrations were a little larger: from 0.02 to 6 mg/L (Figure 7.7).

The analogous behavior of the dependence of the sensitivity on the impurity concentration is observed in this case also. The presence of ammonia impurity at a concentration of about 0.01 mg/L leads to the variation in the film parameters. This is recorded in the 0.1 m absorption spectrum (Figure 7.8).

The use of data obtained from the waveguide spectroscopy methods allows one to examine the mechanism of processes, which are responsible for changes in the optical properties of thin-film waveguides.

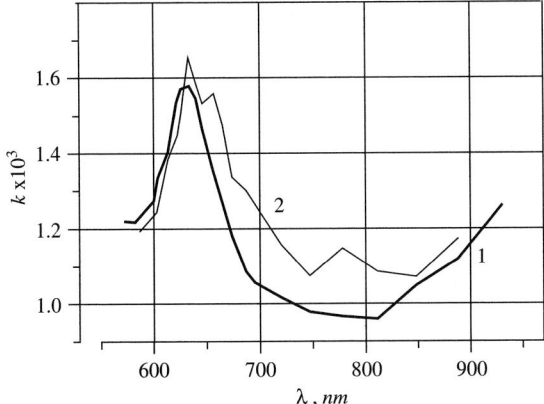

Fig. 7.8. Absorption spectra of SnO_2 films containing antimony oxide in air and air with ammonia admixture (curves 1 and 2, respectively).

7.3. Physical Origin of Processes on Surfaces of Thin-Film Sensors

In the process of creating gas sensor the interaction of semiconductors with the surroundings should be taken into account, because this determines the change in the electrical properties such as the surface conductance of the semiconductors (see Figure 7.7).

To understand the performance of prism-coupler sensors in a qualitative way the following assumptions are made: In the process of gas adsorption we can separate physical adsorption and chemisorption. During the process of physical adsorption the surface "attracts" the parts of gaseous substances to the solid, and this causes growth of gaseous substances near the solid surface. The adsorbed layer is characterized by the absence of chemical bonds between the surface and the molecules (atoms) of gas. The origin of the power causing the attraction of gas molecules to the surface can be explained by van der Waals' interaction. In the case of chemisorption the gas molecule interacts with the surface and creates the chemical bond. Since the sensor is a reversible device, the gas molecules should be are deposited on the surface of the semiconductor in such way that they can leave the surface at any moment. At the same time, in order to change the electrical properties of thin film the following condition should be satisfied. Free-charge carriers in the near-surface region of the semiconductor and adsorbed particles should modify the electron levels of the surface states of the semiconductor. In this case there is a charge exchange between the adsorbed particles and the surface. Thus, sensor functioning is determined by a pronounced chemical adsorption [236]. The physical background of adsorption processes and the phenomenological theory of the variation of a semiconductor film property under the influence of gas impurities are described quite accurately in Refs. [41, 236–238]. We will consider the special properties of gas adsorption on the surface in order to understand the physical background of the processes taking place on the sensor surface.

The surface of the semiconductor is characterized by the presence of a system of energy levels on the surface states distributed in a band gap. If we look at the distribution of these states over a typical band gap we will obtain a diagram similar to that given in Figure 1.4. At room temperature the majority of electrons and holes are close to the band-gap edges but are distributed non-uniformly. The condition of the electroneutrality of semiconductors requires that the band of surface states is to be filled in half [60] and its center is to be near the middle of the band gap. While doping the semiconductor with donor impurity the Fermi level is shifted to the bottom of the conduction band. But it is more energy advantageous for electron to leave the bulk donor levels and move to the free levels of surface states situated below. The Fermi level is decreased as a result of charge transitions and the surface band is filled until the balance state is reached. The uncompensated positive donor ions create an electric field in the depleted layer with thickness d_{SCR}. The Poisson equation for this region is

$$\frac{d^2 U}{dy^2} = -\frac{4\pi e N_d}{\varepsilon}, \qquad (7.3.1)$$

where y is the coordinate in the direction normal to the surface, ε the permittivity of the semiconductor and N_d the concentration of doping impurity in the bulk.

Thereby,

$$U(y) = U_v - \frac{2\pi e N_d}{\varepsilon}(y - d_{SCR})^2,$$

and the change in the electron energy levels near the semiconductor surface caused by this electrostatic potential is referred to as the band bending [41].

Let the gas molecules, e.g., oxygen, be adsorbed on the surface of the semiconductor having the pronounced covalent bond (see Section 1.1). In the oxygen molecule the electrons fill bonding 5σ and 1π orbitals, and two non-shared electrons (Figure 7.9) fill antibonding 2π orbital. The similar scheme depicted in Ref. [41]. Owing to the not large internal bonding energy (5.2 eV), this molecule relatively easily dissociates, since the energy can be

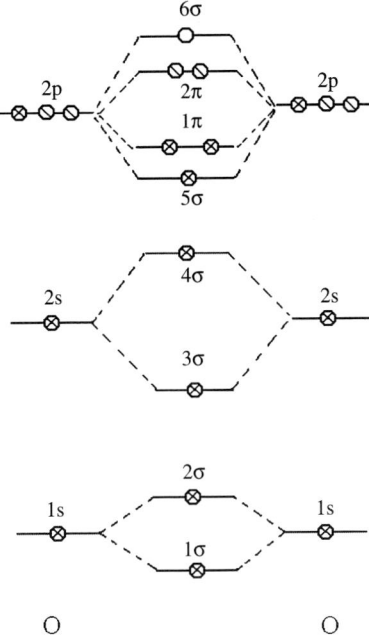

Fig. 7.9. Scheme of orbitals of the oxygen molecule.

compensated by the local bonding between the surface and oxygen atoms [236, 237].

Thus, the chemisorption of oxygen on the surface of the major metals and ion semiconductors occurs, as a rule, with the dissociation of the oxygen molecule [41, 64]. With the excitation of the occupied surface state, it returns an electron obtained from the doping impurity state in the bulk. This electron returns to the bulk, hence the region of surface charge is narrowed and the straightening of bulk bands occurs. The surface state now is ready to form the covalent bond with the oxygen $2p$ orbital oriented normal to the surface. The previous surface state disappears from the band gap and from the new surface state with considerably lower energy, which is close to the energy of the $2p$ orbital of the oxygen. The adsorption behavior is changed for the surface of the semiconductor, where the bonds of ion type prevail, e.g., for the surface of zinc oxygen semiconductor ($E_g = 3.2\,eV$). The experimental data show, that the oxygen is adsorbed on polar surfaces (Figure 7.10) as well as on non-polar surfaces ZnO, which usually is the n-type semiconductor, and also on the surface of many other ion semiconductors and dielectrics such as the O_2^- ion [41, 236]. The similar scheme depicted in Ref. [41].

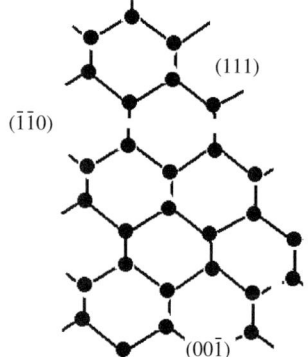

Fig. 7.10. Model explaining the creation of polar and non-polar surfaces.

It is known that the oxygen molecules are oriented along the surface and attract electrons from the semiconductor conduction band, the region of positive space charge being formed in the former location of attracted electrons, i.e. each molecule in the adsorption process should take one bulk electron. Such charge transport increases the width of the depletion layer and consequently the degree of band bending. The increase in the barrier level makes the transport of charges difficult. As a result the rate of adsorption is decreased exponentially.

Following the description given in Ref. [239], let us consider the phenomenological model of changes of the semiconductor film properties owing to the influence of the gas impurities. Chemical adsorption in semiconductor materials is known to change their surface conductivity. The potential in the surface charge region (SCR) is defined as a rule by the Poisson equation:

$$\Delta U = 4\pi\rho/\varepsilon, \tag{7.3.2}$$

where $\rho = e(N_D - N_A + N_p - N_e)$ is the charge density in the SCR caused by the semiconductor-gas contact, N_D and N_A are the concentration of ionized donors and acceptors and N_e and N_p the concentration of free electrons and holes, respectively.

Owing to the total electroneutrality of the semiconductor bulk we can write the following relationship:

$$N_D - N_A + N_{p0} - N_{e0} = 0.$$

Here N_{e0} and N_{p0} are the equilibrium electron and hole concentrations in the semiconductor bulk, respectively.

As the charge carrier concentration is changed only along the direction that is normal to the surface (surface is assumed to be placed in the plane $x = 0$ and semiconductor is placed in the region $x > 0$), the following expression for the Poisson equation can be written for this region in the case if the Bolzman distribution is applied:

$$\frac{d^2 y}{dx^2} = \frac{1}{2L^2}[\gamma(1-e^{-y}) + \gamma(e^y - 1)] \quad (7.3.3)$$

where $y = eU/k_B T$ is the dimensionless potential, $L = \sqrt{(\varepsilon kT)/(8\pi e^2 n_i)}$ the shielding length, $\gamma = N_e/N_i = (N_{e0}/N_{p0})^{1/2} = \exp(\mu/k_B T)$, $N_i = (N_{e0} N_{p0})^{1/2}$ is the electron and hole concentrations in the semiconductor with gas impurity and μ is the chemical potential.

Taking into account the boundary conditions

$$\rho = 0, \ y = 0, \ dy/dx = 0 \quad \text{at } x \to \infty$$

and integrating expression (7.3.3) the following equation can be obtained:

$$\frac{dy}{dx} = L^{-1} F(y, \gamma), \quad (7.3.4)$$

where

$$F(y, \gamma) = \pm \sqrt{(e^{-y} + y - 1)\gamma + (e^{-e} + y^{-1} - 1)\gamma^{-1}}.$$

It should be noted that the function $F(y, \gamma)$ is positive at $y > 0$ and has negative values at $y < 0$. Let us note that y is the value of the dimensionless potential near the semiconductor surface, so the positive values correspond to the energy band bending upward.

The surface density of electrons ΔN_e and holes ΔN_p in the SCR are defined by the expressions

$$\Delta N_e = -N_{e0} L \varphi(y_s, \gamma), \quad \Delta N_p = N_{p0} L \phi(y_s, \gamma),$$

where

$$\varphi(y_s, \gamma) = \int_0^{y_x} \frac{1 - e^{-y}}{F(y, \gamma)} dy$$

$$\phi(y_s, \gamma) = \int_0^{y_s} \frac{e^y - 1}{F(y, \gamma)} dy.$$

The total charge in the SCR is equal to

$$\theta = 2N_i e L F(y_s, \gamma). \quad (7.3.5)$$

The charge in the SCR equals the charge situated on the semiconductor surface, but has the opposite sign; it follows from the total electroneutrality of semiconductor. As this charge depends on the concentration of the adsorbed gas ions, the density of surface charges should also depend on the pressure of the gas in the medium surrounding the semiconductor.

Therefore, the magnitude of charge in the SCR depends on the gas pressure P:

$$Q = -(Q_0 + eN_s(P)) = -Q_s(P). \qquad (7.3.6)$$

Here $N_s(P)$ is the surface concentration of the adsorbed atoms and Q is the surface charge density when adsorption is absent.

Analyzing Eqs. (7.1.5) and (7.1.6), the surface potential can now be considered as a function of the gas pressure. This functional dependence cannot be obtained analytically for a general case. To obtain the analytical expression for the surface conductivity change,

$$\Delta\sigma(P) = e(\Delta N_e \mu_{ns} + \Delta N_p \mu_{ps}),$$

Eqs. (7.3.3) and (7.3.4) should be solved, the analytical solution being derived only in some particular cases. The dependence of the surface conductivity on the pressure of the adsorbed gas can be determined with the help of values of functions $F(y_s, \gamma)$, $\varphi(y_s, \gamma)$, $\Phi(y_s, \gamma)$ for certain values of y_s and γ [239]. In this case the dependence of the adsorbed atom concentration on the gas pressure of the surroundings is determined by the well-known kinetic equation [239]

$$N = N_t \frac{Pb}{Pb+1} = N_i \frac{\psi}{\psi+1}, \qquad (7.3.7)$$

where b is the kinetic absorption constant, $\psi = Pb$ the normalized pressure and N_t the number of absorption surface states.

Taking into account (7.3.6) one can obtain

$$2n_i LF(y_s, \gamma) = N_t \frac{\psi}{\psi+1} - N_0, \qquad (7.3.8)$$

where

$$\psi = \frac{f-1}{\lambda - f + 1}, \quad f = 1 + \lambda \frac{\psi}{\psi+1},$$

$$f = -(2n_i L)/(N_0) F(y_s, \gamma), \quad \lambda = (N_t)/(N_0).$$

Thus, the expressions for the concentration of charge carriers can be written as

$$\Delta N_e = -N_{e0} L\varphi(y_s, \gamma) = \gamma \frac{N_t}{\nu}(y_s, \gamma), \qquad (7.3.9)$$

$$\Delta N_p = N_{p0} L\Phi(y_s, \gamma) = \gamma^{-1} \frac{N_t}{\nu}\Phi(y_s, \gamma) \qquad (7.3.10)$$

where $\nu = (N_t)/(N_i L)$.

The values of ΔN_e, ΔN_p, y and ψ can be found from (7.3.9) and (7.3.10), hence the relationship between the charge carrier concentration and the gas pressure can be derived. For this reason the sign of N_0 in (7.1.7) should be determined. The value of N_0 is negative when the adsorption is absent, and all electrons in the semiconductor are captured by the surface states. But, N_0 is positive, when the holes in the semiconductor are captured by surface states. For example, let us consider N_t, N_0, N_i and L values typical for silicon and consider the case of n-type semiconductor, i.e. the case of positive N_0 and the donor type of gas adsorption taking place on the semiconductor surface, i.e. $F(y_s, \gamma) > 0$. Taking $N_t = 10^{13}$ cm^{-2}, $N_{0Si} = 10^{11}$ cm^{-2}, $n_{iSi} = 1.4 \times 10^{10}$ cm^{-3}, $L = 2.2 \times 10^{-3}$ cm, and $\gamma = 100$ ($N_0 = 10^4 p_0$) as a result, the expression describing dependence of ΔN on the gas pressure can be obtained:

$$\ln |\Delta N(p)| = k \ln \psi. \qquad (7.3.11)$$

This linear dependence is in good agreement with the experimental results (see Figure 7.7, curve 3).

Thus, under conditions of equilibrium, owing to the electroneutrality of the structure as a whole, the electric charge captured by the surface states of thin film is neutralized by the charge of the opposite sign located in the near surface region of the semiconductor film. During the illumination of the semiconductor film the increase in free carrier concentration and deblocking of unsaturated molecule bonds occur. The appearance of the excess carriers shifts the balance between charges in the surface states and in the SCR. This leads to a decrease in the size. During continuous illumination, as in our case, new quasi-equilibrium distribution of carriers in the SCR is formed. The charge exchange between surface states appears in the change of the absorption coefficient of the waveguiding semiconductor layer and causes a change in the imaginary part h'' of the guided mode propagation constant, which is recorded experimentally. These surface states are caused by the adsorbed molecules of the gas with unsaturated molecule bonds and by the allowed energy levels.

Besides, the non-monotone character of the dependence of the propagation constant on the impurity concentration reveals the presence of one more mechanism of the processes taking place in this structure. Thus, in the small concentration range of gas impurities, the adsorption caused by the coordination mechanism is the basic one [235]. During filling of the coordinated-unsaturated centers of near-surface layer there is an increase of the h'' value and a decrease of h'. At larger concentrations the mechanism of hydrogen bonds prevails [235], and the value of h'' reaches a saturation level (Figure 7.6b) and h' begins to increase (Figure 7.6a), this is probably caused by the physical adsorption of the gas molecule on the surface of the semiconductor film.

7.4. Evaluation of the Adlayer Parameters

We can estimate the thickness of the adsorbed layer on the waveguide surface for the given dependence of h' on the concentration gas impurity (Figure 7.6a) [231]. The two mechanisms of the dielectric permittivity variation of the waveguiding structure causing the change in the mode propagation constant should be taken into account (see Section 7.3). We can separate the influence of each of these mechanisms using the fact that starting from some value of the concentration of the gas impurity its further increase does not lead to the significant changes in h'', where the creation of the adsorbed layer has the main influence on the variation of the mode propagation constant.

In order to evaluate the thickness of the adsorbed layers we have to determine the parameters of the waveguiding structure used in the experiment. The thicknesses, $d = 0.1$ μm, $d_g = 0.71$ μm, and the refractive indexes $n = 1.90819$, $n_g = 1.4764$, of the waveguiding film and the buffer layer, respectively, and also the absorption coefficient of film $k = 4.23 \cdot 10^{-4}$, were determined with the help of the mode propagation constant when the structure was placed in a different media, e.g., air and water. The water and air refractive indexes are equal to 1.3328 and 1.0003, respectively. The prism refractive index is equal to $n_p = 1.93601$. The values of ε for bulk materials were taken as the permittivity for adsorbed layers. The results of these estimations are given in Table 7.1, where the settlement changes h' and the adlayer thickness associated with them [231] are provided. They agree well with the experimental data. Here, it is necessary to note that it is also possible to estimate the parameters of the adlayers by measuring the magnitude h' of the guided mode in the case of TE and TM polarization of the incident light [235].

Thus, the use of the integrated-optics sensors allows one to evaluate the parameters of submicron layers deposited on the surface of the waveguide.

Table 7.1. Results of evaluation of adsorbed layer thickness and its influence on parameters of waveguiding structure

	Ammonia	Alcohol	Acetone
ε	1.83603	1.84498	1.83386
d (Å)	3.7	8.5	2.2
$\Delta h'_{calc}$ *	1.0×10^{-4}	1.31×10^{-4}	3.9×10^{-5}

*Calculated values of $\Delta h'_{calc}$ are given for range II (see Figure 7.6a).

7.5. Sensors with Recording of the Light Beam Reflection Coefficient

In the case of environment monitoring and the measurement of impurity concentration in media it is appropriate to use the scheme of the recording of the light beam reflection coefficient when the guided mode is excited by the prism coupler. Modes are directed by the tested waveguiding multilayer structure deposited on the base of the prism coupler. In this case the sensor system becomes simpler and consists of the radiation source, sensing element and the photodetector, which can be placed on the output face of the prism coupler, and the signal-recording device.

7.5.1. Gas Sensors

Similar to the structures based on tin oxide, the structures fabricated by the deposition on the prism coupler ($\varepsilon_p = 3.06145$) of the waveguiding film ($d = 4.98$ μm, $\varepsilon' = 2.3173$), made from acrylic resin, activated with azo-dye "methyl red", can be used as the sensing element of the gas detector. The source of radiation is the single-mode He–Ne laser ($\lambda_o = 632.8$ nm). The width of the Gauss light beam is equal to $a_\omega = 91$ μm. The measurements of the complex propagation constant of the principal mode have been performed already. These parameters were determined using the waveguide method with the device, whose scheme is given in Figure 4.5, and $k_0^{-1}h'|_{C=0} = 1.52072$, $k_0^{-1}h''|_{C=0} = -3.07 \times 10^{-4}$. Now, it is possible to reconstruct the dependencies of h' and h'' on the concentration of the gas impurity by measuring with the sensor placed in the transparent isolated container filled with air and ammonia impurity of the known concentration C. The measurements of the reflection coefficient are usually performed at angles close to the resonant angles (Figure 7.11).

The changes in the real and imaginary parts of the guided mode propagation constant are concerned with the change in the reflection coefficient. The interpolation of the real and the imaginary part dependencies on the gas impurity concentration at $C < 4.3 \times 10^{-4}$ mg/L gives $D_1 = k_0^{-1}\partial(h')/\partial C = -1.32$ L/mg, $D_2 = k_0^{-1}\partial(h'')/\partial C = -0.135$ L/mg, $D = D_1/D_2 = 9.81$. The optimal dependence $R(C)$, which is shown as curve 1 in Figure 7.12, corresponds to the obtained value of D. Curve 1 was calculated on the basis of optimization theory, developed by the co-workers of the institute [2 3 0]. The optimization of the sensor sensitivity for the fabricated waveguiding structure can be performed by increasing the light beamwidth up to $w = 870$ μm and by the appropriate choice of the mode excitation angle. The

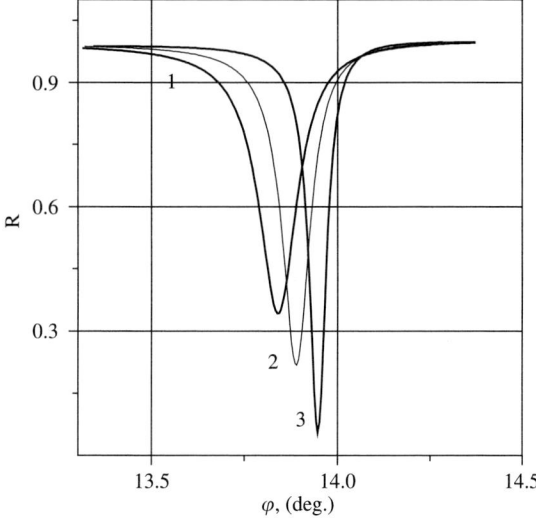

Fig. 7.11. Changes in angular dependence of the reflection coefficient at variations of the ammonia impurity concentration (initial, 40 ppm, 80 ppm – curves 1, 2, 3, respectively).

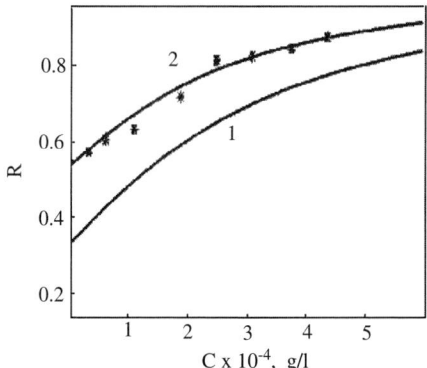

Fig. 7.12. Dependence of the reflection coefficient on ammonia concentration.

$R(C)$ dependence is shown in Figure 7.12 (curve 2), where the discrete points represent the experimental data.

The results of the investigation of the tin dioxide sensor sensitivity to the impurities in the environment are depicted in Figure 7.13. Curve 1 corresponds to the initial distribution of the reflection coefficient, curve 2 its change during the successive gas impurity injection into the analyzed

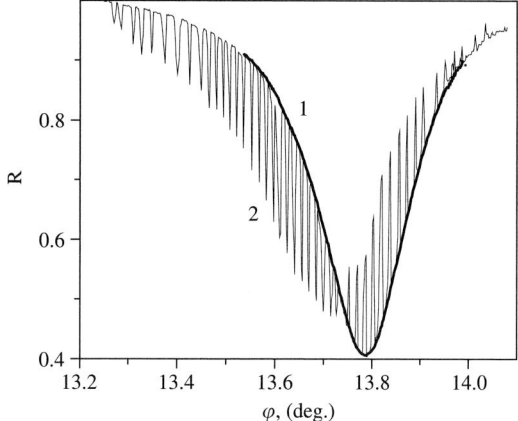

Fig. 7.13. Variations in the sensitivity at the angle shift from the resonant minimum.

container and its further removal from the container. As is evident from the given dependencies, the maximal changes of R are observed, when the angular shift from the resonant angle of the excitation of the guided mode is equal to 0.05° [245].

The time parameters of the sensor, such as operating time and relaxation time, are studied at the sample rate of 0.1 s. The operating time means the time interval required for the creation of the stationary picture in the reflected light beam intensity distribution after injection of the gas impurity.

The relaxation time is evaluated by the 90% level of restoration of the initial parameters of the waveguide after removal of the impurity from the analyzed medium. The operating time is equal to 6–9 s and the time of the full restoration of the initial parameters is 15–20 s (Figure 7.14). The time parameters of the sensor can be improved by decreasing the waveguide thickness.

7.5.2. Integrated-Optics Sensors of the Angular Movement

It is appropriate to consider one more application of the recording scheme of the reflection coefficient of the light beam – the application of the prism coupler with the thin-film waveguide deposited on the prism base in the sensors recording the angular displacements. A number of variations in the schemes of sensing systems based on well-known physical effects, and recording of the linear or angular displacement of the light beam formed by illuminated or luminous moving objects have been developed. Such

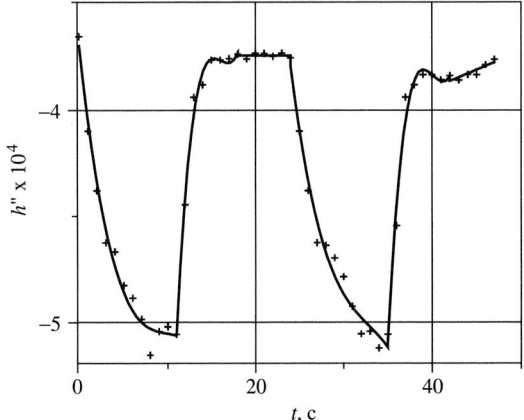

Fig. 7.14. Dynamics of the propagation constant variation at the angle shift from the resonant minimum.

devices based on the principles of interferometry [246] or using the effect of total reflection [247] are quite difficult as they are fabricated on the basis of a multilayered reflecting structures. The sensitivity of the angle measurement can be increased exponentially by increasing the number of the layers [248].

While recording the reflection coefficient during prism excitation of the guided mode the resonant properties of the modes allow one to easily record the angular displacements of the object, with magnitude of the displacement $\simeq 1$ angular second, the reflection coefficient changes being 0.5% (Figure 7.13) and the measurement error -0.1%. Such angular displacement corresponds to the change in the position of the object being of 20 cm and at the distance between the object and the measurement systems equal to 1000 m. The optimization of the prism coupler allows one to increase the sensitivity of the system without any changes in its simple single-layer construction. The use of converging light beams enlarges the range of measured angles, as the width of the minimum in the angular dependence of the light reflection coefficient is increased.

The applications of the devices based on the prism coupler as integrated-optics sensors demonstrate certain advantages, which are specific for this type of resonant devices. It has a large interaction length, significant sensitivity, quite high operating speed and wide range of possibilities of optimization of its not complicated construction. Note that the radiation wavelength can be chosen depending on the spectroscopic peculiarities of the analyzed objects.

7.6. Waveguide Microscopy of Thin Films

High sensitivity of the integrated-optics sensors allows one to use them as the working element of some analog of the microscope. The main aim of microscopy of any type is the recording of correct image of the investigated object. This problem especially occurs while studying of the transparent objects, such as channel waveguides or planar lens in glass. The possibilities of phase-contrast microscopy in this case are limited by insignificant difference in the refractive indices ($\sim 10^{-3}$) of the investigated object and its surroundings. The principles of the waveguide microscopy are simple and similar to the principles of the tunnel microscopy [41, 49]. If one uses the integrated-optics sensor as the sensing element (see Figure 7.5), the parameters are close to the parameters given in Section 7.3, then the field of the guided mode "leaks" outside the waveguiding layer and according to the wave theory can react on the change of the parameters of the environment. Let us decrease the distance between the surfaces of the sensor and the investigated sample in order that the exponential "tail" of the guided mode penetrates into the investigated structure. If the sensor is excited by the wide parallel beam we can observe the image of the investigated transparent sample in the light beam reflected from the base of the sensor (Figure 7.15).

In the scheme, first, it is obvious that any direct observation of the surface gives the possibility to obtain the information about the surface, and second, it imposes the limitations on the models proposed for observation of the obtained results [241]. The model, which is approximate but allows one to interpret images obtained with the waveguide microscopy, was proposed in Ref. [242]. Interpretation of the obtained image on the basis of the proposed model allows one to obtain the distribution of the refractive index in the plane of the sample under test. Let us consider as an example the

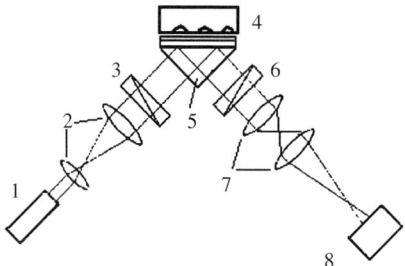

Fig. 7.15. Scheme for observation of phase objects: source of coherent radiation (1), collimator (2), polarizer (3), investigated sample (4), waveguided sensor based on the prism coupler (5), analyzer (6), optical microscope (7), CCD-area imager (8).

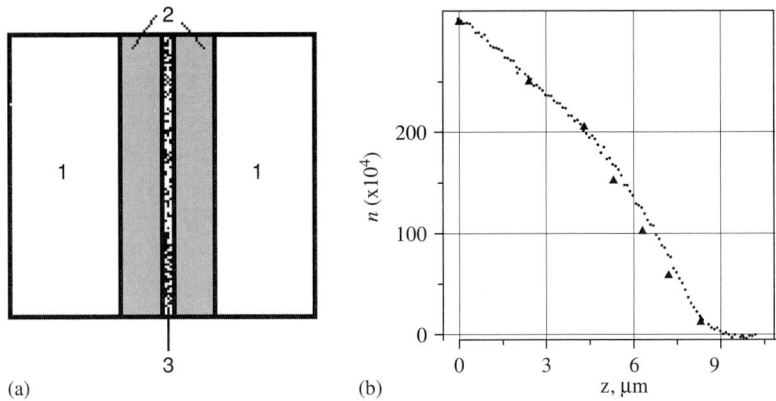

Fig. 7.16. Investigated structure (a): substrate (1), planar waveguide (2), immersion (3 with $n = 1.5150$), and distribution $\varepsilon(z)$ for the investigated structure (b) Continuous curve is data of the waveguide microscope. ▲- Data obtained by WKB method.

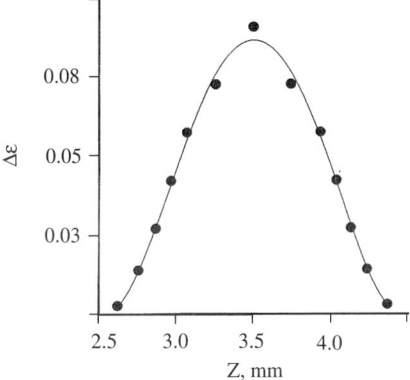

Fig. 7.17. Distribution of the refractive index for the gradient cylindrical lens. •- Data obtained by VKB method.

reconstruction of the refractive index distribution for the structure shown in Figure 7.16a.

The structure was obtained by ion exchange in optical K8 glass. The recorded distribution of the reflection coefficient for this structure allows one to restore the distribution of the refractive index of this planar waveguide (Figure 7.16b). For checking the feasibility of the obtained results the distribution of the refractive index for this waveguide was restored by the WKB method [243]. The obtained profile is also shown in Figure 7.16b. The satisfactory correlation of the data obtained by different methods

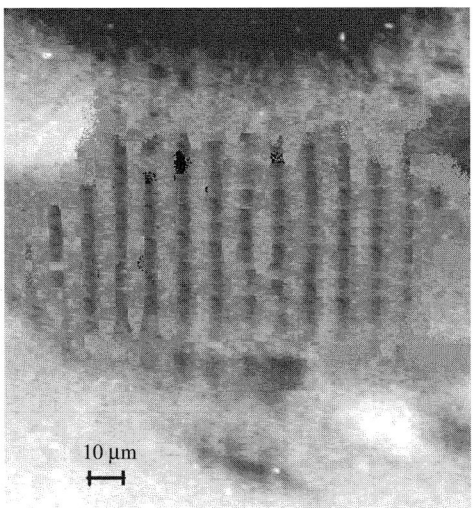

Fig. 7.18. Image of channel waveguides obtained by diffusion of silver into the K8 glass.

allows one to speak about the possibility to apply the suggested method for the determination of the parameters of the optical components. The experimental data testify to the increase in the image contrast during the observation in the direction normal to the smallest size of the object.

The distribution of permittivity in the cross-section of the gradient cylindrical lens was determined by such a setup (Figure 7.17). The image of channel waveguides in optical K8 glass shown in Figure 7.18 is obtained in a similar way. The results of such investigations are given in Ref. [244], where SPW is used as the sensing element for the study of the objects with a size equal to 100 μm.

Thus, the waveguide microscope technique allows one to obtain not only the image of the transparent objects but also the information about the refractive index distribution in the plane of such investigated object as the channel waveguide, gradient cylindrical lens, or gradient blanks of fibers.

Chapter 8
Optical Nonlinearity in Thin Films at Low-Intensity Light

8.1. Optical Nonlinearity and Thin-Film Nonlinear Constant Measurement Techniques 156
8.2. New Method of Determination of Nonlinear Constants of Films. 164
8.3. Nonlinear Optical Properties of Azo-Dye Doped Polymeric Films. 169
8.4. Optical Nonlinearity in Semiconductor Films. 172
    8.4.1. Nonlinear optical properties of As_2S_3 films 173
    8.4.2. Structure and nonlinear properties of ZnSe films 178
8.5. Nonlinear Absorption in Semiconductor-Doped Glasses. 185

Studying light beam propagation in nonlinear thin films is of great interest because of its wide range of possible applications in optoelectronic devices, photonic switching [249–254]. These structures can also serve as models for studying the fundamental aspects of interaction of the electromagnetic field and nonlinear materials, for understanding and explaining the mechanisms of such effects as optical bistability [253,255], nonlinear absorption of light [256], and light propagation in low-dimensional structures [251,257], etc. The development of quite reliable and compact devices of light modulation and devices utilizing optical nonlinearity of thin-film and low-dimensional structures for the control of light by light, [253,254] is the practical value of these researches.

Significant optical nonlinearity is typical for a wide range of organic and non-organic semiconductor materials [257]. As such attention is paid to the creation and development of semiconductor media and structures with nonlinear optical characteristics. Many researches in recent years were dedicated to the study of the optical nonlinearity in semiconductor nanostructures and low-dimensional structures [251,255]. The nonlinear optical parameters of thin-films nanostructures are defined by their geometry and composition and in some cases may exceed nonlinear coefficients for a bulk media [256,257]. The development of thin-film technology is in close association with the development of new methods of their investigation. The effective techniques for studying thin film properties are optical techniques [267,271,274]. During the last decades the creation of new materials on the basis of quantum-dot structures and low-dimension systems, wide applications of thin films in optics and electronics require the improvement of the existing methods of testing and measuring the parameters of thin-film structures and development of new ones. As the high light intensities required for the observation of nonlinear phenomena at the low power of incident radiation are easily achieved in optical waveguides due to their small cross sizes [259], the integrated-optical methods for the study of optical nonlinearity in thin-film structures is of great interest.

In this chapter, the results of the study of optical nonlinearity in thin-film structures when the intensity of incident radiation is below $0.1 \, \text{W/cm}^2$ and the photon energy is less than the optical band gap, are presented.

8.1. Optical Nonlinearity and Thin-Film Nonlinear Constant Measurement Techniques

Only some examples of nonlinear interaction of radiation with the optical materials required for the explanation of the experimental results are considered. For this reason the list of given references contains only the works required for the illustration of observable effects. A more detailed information about these questions can be found in Refs. [276–281].

When the light beam interacts, especially in the case of high-intensity laser beam, with the optical materials all their characteristics are changed. Generally the polarization $\vec{P}$ is the nonlinear function of the field, which completely describes the response of the medium on the field influence. If there are no nonlinear effects the expression for $\vec{P}$ is given by (1.2.9). In the case of nonlinear interaction, but on the assumption of the quite weak field, $\vec{P}$ can be expanded into a power series of the incident field $\vec{E}$. For the field that can be represented as the series of plane monochromatic waves it is possible to write

$$\vec{P} = \vec{P}^{(1)} + \vec{P}^{(2)} + \vec{P}^{(3)} + \cdots, \tag{8.1.1}$$

where

$$\vec{P}^{(1)} = \chi^{(1)}\vec{E}, \quad \vec{P}^{(2)} = \chi^{(2)} : \vec{E}\vec{E}, \quad \vec{P}^{(3)} = \chi^{(3)} : \vec{E}\vec{E}\vec{E}, \ldots .$$

In expression (1.2.10), $\chi = \chi^{(1)}$ is the linear susceptibility of the medium. If the linear and nonlinear susceptibilities $\chi^{(n)}$ for the medium are known, then it is possible to predict all nonlinear optical effects by the Maxwell equations. In the presence of the external field, the dielectric permittivity of the media is the function of $\vec{E}$. At small $\vec{E}$ the dielectric permittivity $\varepsilon(\omega, \vec{E})$ can also be expanded into the power series of the incident field $\vec{E}$:

$$\varepsilon(\omega, \vec{E}) = \varepsilon^{(0)} + \varepsilon^{(1)}\vec{E} + \varepsilon^{(2)}\vec{E}\vec{E} + \cdots, \tag{8.1.2}$$

where $\varepsilon^{(0)}$ is the dielectric permittivity at $\vec{E} = 0$.

From expression (8.1.2) taking into account (1.2.4) and (8.1.1) the expression for $\varepsilon^{(i)}$ can be obtained as

$$\varepsilon^{(1)} = \chi^{(2)}\vec{E}, \quad \varepsilon^{(2)} = \chi^{(3)}\vec{E}\vec{E}. \tag{8.1.3}$$

As $\chi^{(2)} = 0$ for the crystals which do not have central symmetry, and the nonlinear effects of the fourth order ($\chi^{(4)}$) at light intensities used are insignificant, so for the case of non-absorbing medium one can obtain

$$\Delta\varepsilon = \varepsilon - \varepsilon^{(0)} = \chi^{(3)}|\vec{E}|^2.$$

Taking into account (1.2.9), the nonlinear change of the refractive index is equal to

$$\Delta n = \frac{1}{2n}\chi^{(3)}|\vec{E}|^2. \tag{8.1.4}$$

But, for the so-called Kerr media (only $\chi^{(3)}$ is not equal to zero from the set $\chi^{(n)}$) the relationship of the refractive index and the light intensity I is given by the expression

$$n = n_0 + n_2 I, \tag{8.1.5}$$

where n_0 is the refractive index of the media determined for the low-intensity light (the linear refractive index) [253] and n_2 is the nonlinear optical constant of the material or the nonlinear refractive index.

Nonlinear optical constant has been determined for a wide range of optical materials and at different mechanisms of interaction between the radiation and substance [250]. Thus the relation between nonlinear refractive index and nonlinear susceptibility can be found from (8.1.4) and (8.1.5) taking into account (1.2.10),

$$n_2 = \frac{1}{2n}\chi^{(3)}. \tag{8.1.6}$$

Generally the susceptibility and the refractive index are complex quantities. In the case of the propagation of high-intensity light in the absorbing optical media the behavior of light attenuation is not described by the Buger–Lambert law. The absorption coefficient is not constant and also depends on the light intensity. There is the following attenuation law [266, Chapter 1]:

$$\frac{dI}{dz} = -\alpha I - \beta I^2 - \gamma I^3 \cdots, \tag{8.1.7}$$

where β and γ are nonlinear absorption coefficients.

At low-intensity light the change of the absorption coefficient is proportional to I or $|\vec{E}|^2$, For this reason such processes can be considered as nonlinear optical effects of the third order by the electric field [260]. The nonlinear absorption coefficient k_2 ($\beta = 2k_0k_2$) can be introduced in the same way as the nonlinear refractive index n_2. The absorption coefficient k_2 will describe the field influence on the absorbing media.

Here, it should be noted that in some cases the thin-film structures, especially made from semiconductor materials, are anisotropic. However, because in such structures the optical axis is oriented in the directions, that are normal to the substrate and the direction of the light propagation, the use of nonlinear parameters introduced above is quite correct.

Nonlinear effects are usually observed at significant light intensities. The fact that nonlinear optical effects are usually the wave effects, they should be taken into account. Thus, at high coherence of the incident radiation the accumulation and occurrence even of weak nonlinear optical effects are also possible. Under favorable conditions one can observe nonlinear optical effects at the light intensity of $10\,\text{mW/cm}^2$ [262]. Nonlinear effects of the third order (optical characteristic changes are described by the nonlinear susceptibility $\chi^{(3)}$) are characterized by the dependence of the refractive index on the light intensity I according to the expression (8.1.5). The first experiments on the demonstration of one of such effects is the self-focusing of the light beam ($n_2 > 0$), is related to the dielectric media, which has the value of n_2 of 10^{-6}–$10^{-13}\,\text{cm}^2/\text{W}$. The nonlinear refractive index n_2 was determined for several semiconductors (for example, GaAs: $-n_2 = -4 \times 10^{-4}\,\text{cm}^2/\text{W}$; InSb: $-3 \times 10^{-3}\,\text{cm}^2/\text{W}$). These values are much greater than the values for the traditional nonlinear materials. Other media with "giant" nonlinearity, such as the organic dye, liquid crystals, etc. are found. But semiconductors are still the most attractive for practical applications due to its quick response and the possibility to create miniature devices. The investigations of quantum-dimensional structures [265] and low-dimensional systems are of greater interest because of a wide range of possible applications of these structures in optoelectronic devices [266,252].

The changes in the spectrum-optical characteristics of semiconductors caused by high-intensity laser radiation take place in consequence of such processes as the generation of charge carriers during light absorption, appearance of electric field due to the space charge, the change of the polarization of the crystal lattice at the capture of the charge carrier and others [267,268]. All mechanisms of optical nonlinearity may manifest themselves simultaneously at the illumination of the semiconductor by the laser source, but usually one or two mechanisms are basic [269]. In all cases the effect of these mechanisms appears as the dependence of the refractive index or absorption coefficient on the light intensity. The refractive index is correlated with the absorption coefficient spectrum by the Kramers–Kroning relation [10]. If it is possible to change the absorption spectrum by changing light intensity, the refractive index becomes dependent on the light intensity

(Figure 8.1). In this case the media is a nonlinear substance. The changes in the absorption spectrum lead to variations of the Re ε spectrum. Therefore, the changes in the absorption coefficient $\Delta\alpha(\hbar\omega)$ and the refractive index depend on each other [289]:

$$\Delta n = \frac{\pi c}{\pi} \int_0^\infty \frac{\Delta\alpha(\hbar\omega')}{(\hbar\omega')^2 - (\hbar\omega)^2} d(\hbar\omega'). \tag{8.1.8}$$

The origin of optical nonlinearity can be explained by using the band-filling model. This effect is demonstrated in Figure 8.2 [270]. If the light beam intensity (at $\hbar\omega > E_g$) is high enough to fill the lower power levels of the electron states in the conduction band faster than they decay, the width of the band gap E_g is increased to the $\hbar\omega$ value. In the absorption spectra (see Figure 8.2) one can observe the short-wave shift (the so-called "blue" shift) of the absorption edge [271]. The saturation of absorption during the band filling is known as the Burnstein–Moss dynamic effect [272]. This model can be applied to the "band–band", "band-impurity level" or "band-surface state" transitions. The fact that interband transitions make the basic contribution should be mentioned here. The appearance of free carriers in the conductivity band leads to changes in the refractive index and also in the absorption coefficient [150]. This causes the phenomenon light absorption in the media [40,129].

Every optical transition from the valence band to the conductivity band contributes to the refractive index (see section 1.1). If one takes approximate

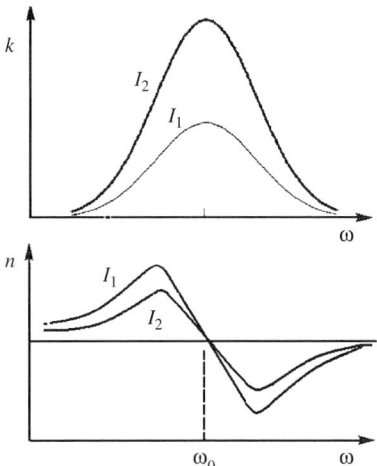

Fig. 8.1. Relationship between the absorption coefficient and the refractive index at the radiation wavelength variations.

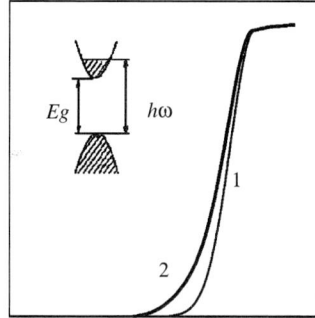

Fig. 8.2. Scheme of the band-filling effect and the absorption edge shift (curve 2) for semiconductors with a direct band gap, basic spectrum (curve 1).

linear relation between the change of the refractive index n and the carrier concentration N as

$$\Delta n = \sigma_a N,$$

where σ_a is the constant of proportionality, $N = \alpha I \tau / \hbar \omega$, τ the lifetime of non-equilibrium carriers and α the absorption coefficient [273], then taking into account (8.1.5) one can write

$$n_2 = \eta \sigma_a \alpha \tau / \hbar \omega, \tag{8.1.9}$$

here η is the coefficient, characterizing the effectiveness of the internal photoeffect.

While investigating the optical characteristics of semiconductor materials near the fundamental absorption edge one can observe the appearance of the excited mechanism responsible for the existence of optical nonlinearity in semiconductors [275]. A more detailed information on this topic can be found in Refs. [276–278].

The appearance of quantum-dimensional effects and the surface or interface effects is another mechanism that leads to the optical nonlinearity in the thin-film structures [250,251,265,277,279–282]. The quantum-dimensional effect appears in polycrystal films or specially fabricated diphasic structures, which contain nanocrystals incorporated into the matrix of basic materials. The corpuscle size is equal to several nanometers (usually of order 10 nm). Quantum-dimensional effects appear when the size of nanocrystal is possible to be compared with exiton Bohr radius. In this case, the wave functions of the charge carriers become restricted by the semiconductor nanocrystal size. This phenomenon takes place in practice in the "blue" shift of the basic absorption edge with the decrease in the nanocrystal sizes. In direct band-gap semiconductors such as II–VI or IV–VI the absorption spectrum near its edge is in the form series of a discrete lines [251,277].

Except the quantum-dimensional effects a significant role in such structures is played by the surface states. This is due to the fact that a significant number of atoms is situated on the surface of semiconductor crystals with sizes of 5–10 nm. This leads to a substantial influence of interfaces on the optical characteristics of thin-film structures. The effect of interfaces can be explained on the argumentation of the following model. Quite often there are deep surface states on the semiconductor surface. The energy levels of these states are situated not far from the middle of the band gap. The density of these states may reach 10^{13}–10^{15} cm^{-2}. Surface states may be of p-type (acceptors) or n-type (donors) (Figure 8.3). The capture of the basic carriers by these states leads to the surface bending of the zone and to the creation of the depleted layer near the semiconductor surface. The thickness of the depleted layer

$$L = \sqrt{\varepsilon e U / 2\pi N e^2}, \qquad (8.1.10)$$

where ε is the permittivity and $eU \approx E_g/2$ is the surface level position relative to the Fermi level, and at $N = 10^{18}$ it is $\simeq 500$ Å [283].

Surface states set the Fermi level near the middle of the band gap [60]. Let us note the fact that the density of the surface states at the interface of the "film-substrate" is little less (10^{12} cm^{-2}) and the lifetime of carriers τ in this case is $< 10^{-12}$ s, while $\tau \leqslant 10^{-6}$ s. at the film surface [283]. The surface states are separated from the bulk states by the potential barrier that decreases the probability of exchange by the charge carriers between the semiconductor bulk and its surface. The magnitude of this barrier is approximately equal to the band incurvature. With the increase in light intensity that irradiates the semiconductor there is a decrease in the band incurvature near the surface as a result of the spatial separation of the charges. It should be mentioned that

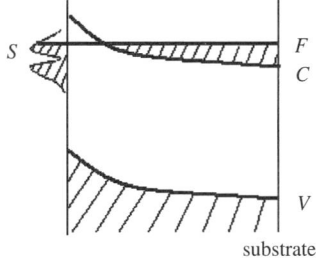

Fig. 8.3. Scheme of band structure for the semiconductor film: Fermi level (F), bottom of the conductivity band (C) and top valence band (V), respectively, surface states (S) semiconductor of n-type.

at a film thickness < 1000 Å the full separation of charge carriers takes place [283]. We can draw the conclusion that at light absorption with $\hbar\omega < E_g$, there is the optical excitation of surface states or impurity levels in the area of spatial charge, these states become long-lived in the area of the band incurvature [284]. The presence of the dangling bonds or impurity atoms on the surface leads to changes in the energy levels of the surface states. Therefore, the modification of the semiconductor surface allows one to create structures with new optical parameters [281].

Nonlinear optical effects may appear sometimes as a result of structural, phase or orientation mechanisms. Such processes are possible in liquid crystals [285] and polymeric materials [249,254,286]. Let us consider the examples of such mechanisms on liquid crystal materials. Typically, these materials consist of long molecules orientated in a random way, they being aligned along the common axis (in some materials separate molecule chains are formed) at the application of external field. But the intensity of the light wave $E(\omega)$ changes at a high frequency. This occurs too fast and the molecules do not follow these changes, and therefore the orientation effects are proportional to E^2. Molecular regulating under the influence of optical wave appears in induced anisotropy Δn of the refractive index n:

$$\Delta n = 2\pi\chi^{(3)} \frac{|E|^2}{n}. \qquad (8.1.11)$$

We have considered some mechanisms of optical nonlinearity and we will return to them while discussing the experimental results. One can get significant information about the source and reason of the medium of optical nonlinearity from the measurements of the nonlinear refractive index. Nowadays, there are a great number of devices and methods of determining nonlinear parameters of the medium. We may mention the methods of self-focusing (defocusing) and z scanning [278–288]. With the help of these methods one can measure the field distribution in the far zone depending on the nonlinear medium position relative to the focal plane of the lens (Figure 8.4). Actually, we measure the change of effective nonlinear focus distance

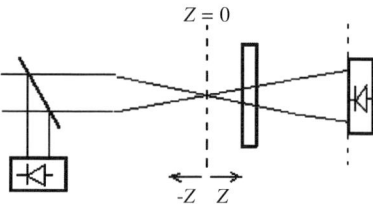

Fig. 8.4. Setup for the observation of optical nonlinearity with the help the z-scanning method.

caused by the focusing effect (defocusing) of the light beam in the nonlinear material under the conditions of self-effect. For example, in the case of cubic nonlinearity, when

$$n = n_o + \frac{n_2}{2}|E_0|^2 = n_o + \Delta n, \quad (8.1.12)$$

the nonlinear change in the electric field phase $\Delta\varphi$ for the Gaussian light beam propagation through the nonlinear medium will be

$$\frac{d\Delta\varphi}{dz} = \frac{2\pi}{\lambda}\Delta n,$$

and the phase shift at the output will be

$$\Delta\varphi(z,z,t) = \frac{\Delta\varphi_o}{1+z^2/z_o^2}\exp\left(\frac{2r^2}{a_2}\right), \quad (8.1.13)$$

where

$$\Delta\varphi_o = \frac{2\pi}{\lambda}\Delta n_o \frac{1-e^{-\alpha L}}{\alpha},$$

$$a^2(z) = a_o^2\left(1+\frac{z^2}{z_o^2}\right), \quad (8.1.14)$$

$$z_o = (\pi/\lambda)a_o^2/2,$$

in which L is the sample thickness, α the absorption coefficient, a the beamwidth and a_0 is the value of a at $Z=0$ [287].

During the change in the range from $Z<0$ to $Z>0$, for example, for the medium with $n_2<0$, firstly the widening of the light beam (maximal beamwidth at $Z=0$) takes place as a result of self-defocusing effect and then the narrowing of the beam occurs. For media with $n_2>0$ the process of the change in the light beam size in plane D_2 has the opposite behavior. While observing these changes one can determine the n_2 coefficient sign and by using expressions (8.1.13) and (8.1.14) one can calculate its value. The n_2 measurement technique based on the deviation of the light beam in nonlinear medium is also worthy of noting [289]. As a result of self-interaction the deviation angle θ_n of the beam measured in the far zone (Figure 8.5) is related to the thickness L of the investigated sample and beamwidth a by the following expression:

$$\theta_n = n_2 LI/a. \quad (8.1.15)$$

This method allows one to calculate the n_2 value quite easily.

There are other methods of nonlinear constant measurement which are more complicated in technical performance. That is, for example, the

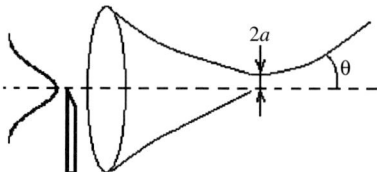

Fig. 8.5. Setup for the study for optical nonlinearity using the self-deviation technique.

method of degenerate four waved mixing (DFWM), based on the investigation of the diffraction of the probe light on the photoinduced dynamic gratings [290,297] or electrooptical constant measurement [291]. These and other methods allow one to determine the nonlinear constants for different wavelengths. Among the methods of detection and investigation one of the most prevalent is the two-beam method. The main feature of this method is the measurement of the absorption spectrum by low-intensity radiation in the absence and in the presence of the excitation radiation with high intensity [292]. The changes in transmission are recorded by the direct measurement of light transmission. This method allows one to investigate the nonlinear absorption spectra at the variation of the intensity of the excitation light [293] or time parameters of nonlinear absorption using the light impulses [294]. Measurements of nonlinear constant n_2 are quite often performed using the interferometry methods [295,296]. One can find the refractive index variation with the help of these methods by the shift of the interference fringes for parallel-sided samples [295] or by the change in the waveguide parameter [296].

All these methods allow one to correctly determine the nonlinear constants of optical materials. But these methods mostly measure only the nonlinear refractive index and require high-intensity laser beams.

8.2. New Method of Determination of Nonlinear Constants of Films

The Fourier spectroscopy of guided modes of nonlinear thin-film waveguides is the result of the development of the measurement techniques used for the determination of thin-film parameters. It is also based on the recording of the spatial intensity distribution of the light beam reflected from the prism coupler of tunnel excitation of nonlinear medium guiding the optical modes.

Recorded changes in the spatial intensity distribution of the reflected light beam take place at the film refractive index variations of the order 10^{-6}.

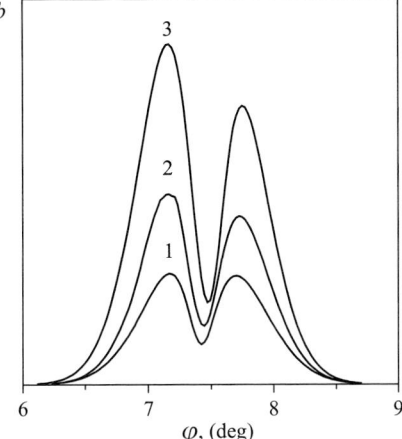

Fig. 8.6. Changes in the angular Fourier spectrum of the light beam in the self-effect case at different light intensities ($I = I_0$ (1), $I = I_1$ (2), $I = I_2$ (3), $I_2 > I_1 > I_0$).

Since these variations can be easily obtained in media with nonlinear refraction there is the possibility of utilization of the technique described in Chapter 4. This approach is based on the recording of the transformation of the Fourier spectrum of the reflected light beam in the case of self-effect [170,298]. At the reflection of the light beam from the nonlinear material in the case of the gradual increase in the incident light beam intensity I its Fourier spectrum changes: minimum angular location φ_o of the spatial distribution and the magnitude of minimum $b_{\min}(\varphi_o)$ of the recorded distribution $b(\varphi)$ are varied (Figure 8.6). The prism coupler contacting the investigated surface is shown in Figure 4.7. The prism with the refractive index n_p is separated from the guiding film with the refractive index n_f by the buffer layer with the refractive index n_g and thickness d_g. The axis of the light beam with the beamwidth a_w, focused on the prism base, coincides with the beam shown in the figure.

The mathematical description of the optical mode excitation in the nonlinear waveguide gives a simple solution that allows one to associate the recorded parameters of the angular distribution of the reflected light beam intensity with the nonlinear parameter P_3 of the waveguiding structure [299]:

$$\frac{\Delta\varphi}{\Delta(\Pi_o^{-1})} = \frac{[S_1 - (2P_2 + P_1)S_2]P_1^2 P_2^2 \, \text{Re} \, P_3}{8P_2(P_1 + P_2) + P_1^2(2P_2 + P_1)^2}, \qquad (8.2.1)$$

$$\frac{b_{\min}(\varphi_0)}{b_0} = \frac{cI}{I_o} + (1-c)\frac{I^2}{I_o^2}, \qquad (8.2.2)$$

where P_1 and P_2 are determined at the intensity I_0 by the results of the processing of the Fourier spectrum of the reflected light beam, the nonlinear effects being not observed at I_0 yet (Figure 8.6, curve 1).

In this case, parameters P_1 and P_2 are expressed as

$$P_1 = \frac{a \mathrm{Im} \hbar}{\sin \alpha}, \quad P_2 = \frac{2k_p k_g |\Delta h a|}{(k_p^2 + k_g^2) \sin \alpha}, \quad (8.2.3)$$

where

$$S_1 = \int_{-\infty}^{\infty} j \, d\xi, \quad S_2 = \int_{-\infty}^{\infty} \xi j_2 \, d\xi,$$

$$j = \frac{\pi}{2\sqrt{3}} \exp\left(\frac{3P_1^2}{4}\right) \left[\exp(P_1 \xi)\left[1 + erf\left(\xi + \frac{P_1}{2}\right)\right]\right]^3,$$

$$\xi = x a_x^{-1},$$

a_x is the light beam projection on the X-axis.

The dependencies $\varphi(I/I_0)$ and $b_{\min}/b_0 = f(I/I_0)$ depicted in Figure 8.7 can be derived from the results of the processing of the spatial Fourier spectra represented in Figure 8.6. The magnitude b_0 is the value of b in the intensity distribution at $I = I_0$. The processing of such curves using

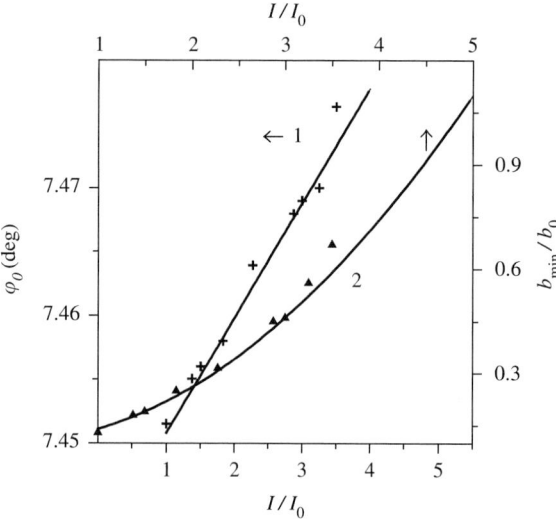

Fig. 8.7. Dependence of the φ_0 and $b_{\min}/b_0$ values on the relative intensity: (▲,+), experimental data; calculated curves [299] are depicted as a continuous line.

expressions (8.2.1)–(8.2.2) allows one to determine the nonlinear complex parameter P_3 of the waveguiding structure:

$$P_3 = \frac{16\sqrt{\varepsilon_a \varepsilon_p} I a \mu_o k_o^2 G}{(a^{-1} a_o \sqrt{\varepsilon_a} + \sqrt{\varepsilon_p})^2 \pi (h')^2 \sin \alpha}, \quad G = \frac{\int_{-\infty}^{-g} \varepsilon_2 |Y|^2 Y^2 \, dy}{\left| \int_{-\infty}^{\infty} Y^2 dy \right| \int_{-\infty}^{\infty} Y^2 \, dy}, \quad (8.2.4)$$

where

$$k_g = \sqrt{(h')^2 - k_o \varepsilon_g},$$

$$k_p = \sqrt{k_o^2 \varepsilon_p - (h')^2},$$

$$h = h' + ih'', \quad \Delta h = \bar{h} - h,$$

$$aa_o^{-1} = \frac{1}{\cos \varphi_o} \sqrt{1 - \frac{\varepsilon_a}{\varepsilon_p} \sin^2 \varphi_o},$$

$$\bar{h} = k_o n_p \sin \left(\theta - \arcsin \left(\sqrt{\frac{\varepsilon_a}{\varepsilon_p}} \sin \varphi_o \right) \right),$$

$$c = 1/[1 + S_1 P_2^2 I_m P_3 / (2P_2 + P_1)],$$

in which ε_2 is the nonlinear complex permittivity of the thin film and $Y(y)$ the field of the guided mode when the incident light intensity $I = I_0$.

One can find the nonlinear refractive index n_2 and the nonlinear absorption coefficient k_2 of the investigated structure material if the refractive index of the film, its thickness and the field of guided mode were calculated previously.

The method, very close in technical essence, of measuring nonlinear optical constants of thin films is proposed by Rignennett et al. [300]. This approach is based on the far-field recording of the changes of the light beam reflected from the prism coupler. However, like all earlier developed integtated-optics methods of measurement of thin-film parameters, this approach allows one to determine only the nonlinear refractive index.

The scheme of the device for measurements of the light beam Fourier spectrum is depicted in Figure 8.8. The measurement technique is similar to the one described in Chapter 4, therefore we will not consider it in details here. Let us note that the block of the light beam intensity measurement and attenuator inserted into the scheme additionally allow one to control the intensity of the probe beam. The problem of control of the incident beam intensity change can also be solved in another way. If we change the position of the focusing element (6) as shown in Figure 8.8 relative to the prism coupling element, the spectrum of the incident beam distribution will remain

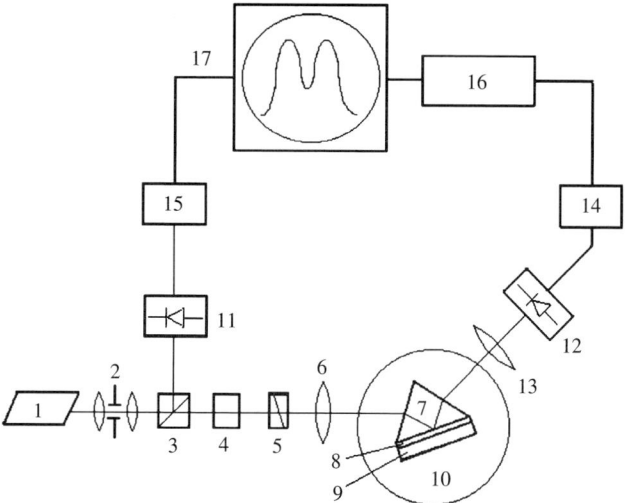

Fig. 8.8. Experimental setup used for determination of optical parameters of thin-film waveguides: light source (1), collimator (2), attenuator (3), beam splitter (4), polarizer (5), lenses (6, 13), prism coupler (7), gap (8), waveguide (9), rotary table (10), photodetectors (11, 12), intensity measurement devices (14, 15), analog–digital converter (16), computer (17).

the same but the intensity will be changed at the prism base according to the changes in the beam radius at the prism base (see expression 8.1.15).

In practice, quite often one needs to evaluate the parameters of the medium nonlinear . There is an easier approach for this:

$$n_2 = \frac{\sin\alpha}{I_o} \frac{\partial(\Delta\varphi)}{\partial(I/I_o)} \frac{a_o d_f (n_a + n_p)^2 n_b}{4 n_a n_p n_f P_2 G^2 \exp(P_1^2/2) J_{\max}}, \qquad (8.21)$$

where

$$n_b = \bar{h}/k_o, \quad G = G_1/G_2,$$

$$G_1 = \frac{1}{2\kappa}[\sin 2\delta - \sin 2(\delta - \kappa k_o d)] + k_o d + \frac{1}{\sigma}\cos^2\delta,$$

$$G_2 = \frac{1}{2}\left[\cos^2(\delta - \kappa k_o d)\frac{1}{\gamma} + G_1\right],$$

$$\gamma = \sqrt{n_b^2 - n_s^2}, \quad \sigma = \sqrt{n_b^2 - n_a^2}, \quad \kappa = \sqrt{n_f^2 - n_b^2}, \quad \delta = \arctan(\sigma/\kappa).$$

Note that owing to the used approximation of the uniform distribution of the intensity over the beam cross-section this approach gives the value of the nonlinear refractive index underestimated by 1.5 times.

8.3. Nonlinear Optical Properties of Azo-Dye Doped Polymeric Films

The capabilities of the stated method for simultaneous determination of the nonlinear constants n_2 and k_2 of the material under the conditions of self action (i.e. with using one light beam) can be demonstrated by the example of thin-film polymeric azo-dye-doped waveguides.

Polymeric films as well as model structures have been chosen as the object of investigation because of the simplicity of their fabrication. Waveguides are usually formed from PMMA solutions and methyl red dye in the mixture of chlorobenzene and dichlorethane dissolvents by the technique of centrifugation [300] or by the technique of chemical diffusion of the dye [301] into multicomponent dissolvent based on the xylol. Propagation losses measured at the wavelength $\lambda = 0.633$ nm are <1.5 dB/cm in single-mode waveguides, and of 3–4 dB/cm in multimode waveguides. Profiles of the guiding layer refractive index, calculated by the WKB method [243], were similar to the step-like profiles. The value of the refractive index increment Δn on the surface of the waveguide obtained from the same dissolvent does not depend on diffusion time t, and the waveguide thickness d is proportional to $t^{1/2}$. The increase in the concentration of the xylol in the dissolvent, from 20% to 40%, leads to the increase in Δn approximately two times, but the losses rise significantly due to the deterioration of the optical property of the waveguide surface. Analogous tendencies occur with the increase in the dissolvent temperature. Maximum obtained values of Δn were 0.012 at an acceptable loss level, the concentration of the dye being of 0.04 Mol/L. The absorption spectra of red methyl dye in the obtained diffusive layers do not differ from the spectra of its many-dimension trance isomer in other polymeric hosts [302]. The spectra of induced absorption under the influence of linear polarized light at the wavelength $\lambda = 515$ nm (Figure 8.9, curve 1), which were measured by the photomodulation technique at the frequency 4 Hz, are also analogous to the spectra investigated before [303].

In the 415–565 nm spectral range the samples intensively became transparent, but in the yellow–red spectral range they became slightly opaque. Only the short-wave band was essentially dichroic. The sensitivity of the investigated layers to the light at $\lambda = 633$ nm was also discovered in such materials. The changes in the absorption spectrum MR (Figure 8.9, curve 3) caused by the light are opposite to the changes observed under the influence of short-wavelength radiation, only the "red" band being dichroic. In all cases the sample becomes transparent in the spectral range around the wavelength of the illuminating beam. The fact is that the additional illumination of the sample by unmodulated radiation at $\lambda = 515$ nm with comparable

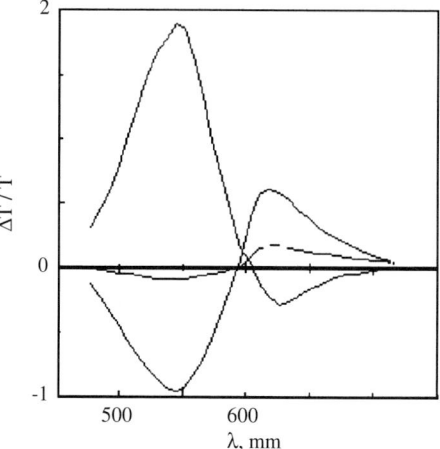

Fig. 8.9. Spectra of the photoinduced change of the transition coefficient of the polymeric film doped by MR, at the light wavelength $\lambda = 515$ nm (1, 2) and 633 nm (3). Curves 2 and 3 are obtained under additional illumination at $\lambda = 633$ (2) and 515 nm (3). The intensity of green light in all the cases is equal to $5\,\text{mW/cm}^2$, and that of red is $100\,\text{mW/cm}^2$ [303].

intensity makes the effect stronger. The sensitivity of the layer to the green light is also increased, when the sample is illuminated at the same time by the red light (Figure 8.9, curve 2). In both the cases the magnitude of the response almost does not depend on the polarization of the third light beam.

Let us consider in more detail the procedure of measuring the nonlinear constants n_2 and k_2 by utilizing the method described in Section 8.2 and performed with the help of the experimental setup depicted in Figure 8.10. When the incident light intensity $I_o \approx 0.5\,\text{mW/cm}^2$, the linear parameters of the waveguiding film are determined from the results of the processing of the measured Fourier spectra of the reflected light beam in the case of excitation of the guided modes. These parameters are equal to $n = 1.5221$, $k = 70.9\,\text{cm}^{-1}$ and the film thickness $d = 3\,\mu\text{m}$. During the consecutive increase in the incident light intensity in the range from $0.5\,\text{mW/cm}^2$ to $0.1\,\text{W/cm}^2$ it is necessary to construct dependencies $\varphi_{\min}(I/I_0)$ and $b_{\min}/b_0(I/I_0)$. Thereby, one can determine n_2 and k_2 for the studied structures by processing these dependencies. Thus, the photoinduced decrease in the refractive index and the induced bleaching of the layers under the influence of the light propagated in them were recorded. The dependence of the nonlinear constants on the concentration of azo-dye MR are shown in Figure 8.10, where $n_2 = \Delta n/\Delta I$, $k_2 = \Delta k/\Delta I$. The saturation of n_2 in the case of the azo-dye concentration increase is probably caused by the shift in the low-frequency spectrum range of the absorption band having the maximum at 610 nm

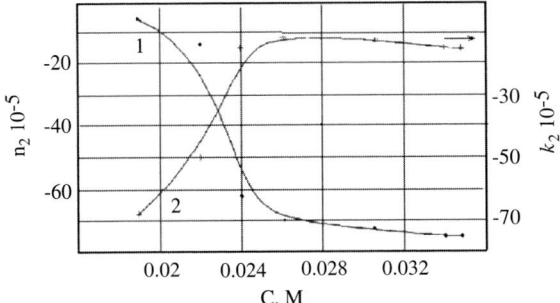

Fig. 8.10. Dependencies of nonlinear parameters n_2 (1) and k_2 (2) on azo-dye concentration for films doped by azo-dye MR.

(Figure 8.9). The stabilization of k_2 observed during the increase in the dye concentration C is caused by the presence of at least two mechanisms of optical nonlinearity in such films. This requires further investigation.

An explanation of the observable phenomena can be given considering the possibility of the inverse process of the photoinduced *cis–trans* isomerization not only by thermal activation, but also with the illumination by of the red light, i.e. in the presence of the photochromic properties of *ci* isomers of MR. As it follows from Refs. [301,307] the *trans* isomer of MR absorbs light at $\gamma \leqslant 565$ nm, and *cis* isomer at the light wavelength equal to 400 nm and lower. The low-frequency band of the induced absorption in photomodulation spectra 1 and 2 (Figure 8.9) relate, obviously, to $n - \pi^*$ transition in *cis* isomers [304].

When the absorption of light is of 633 nm wavelength the equilibrium concentration of *cis* isomers is decreased, and the sample becomes transparent in this spectral range. Thus, owing to the increase in the part of the *trans* isomers of MR the absorption is increased in the band with the maximum at 495 nm. The fact that these bands of induced absorption are not simultaneously dichroic also specifies them belonging to different azo-dye isomers.

Thus, the additional illumination changes the dynamic balance in the system and in the concentration of the *cis* or *trans* isomers of MR dye, respectively. The sensitivity of the investigated samples only to the red light means that certain portion of the dye is in the *cis* form. The confirmation to this fact can be found in the absorption spectra [303] of such films (Figure 8.11), measured by the waveguide method described in Chapter 5, with various polarizations of the incident light (TE and TM). The spectral dependence of the absorption coefficient of this film in the visible range is shown in Figure 6.7.

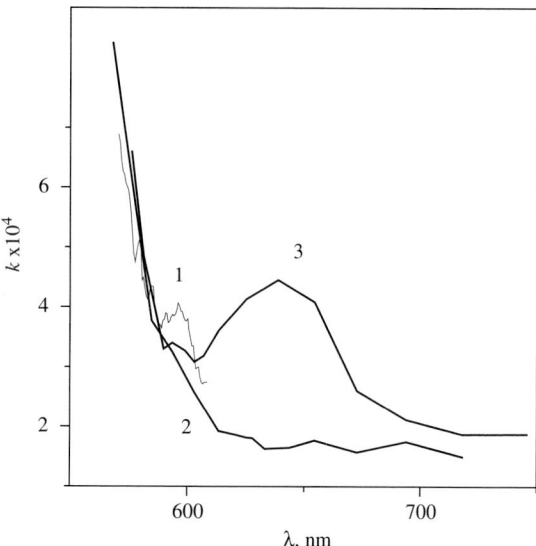

Fig. 8.11. Absorption spectra of the MR-doped PMMA film measured by the transmission photometry method (1) and the waveguide spectroscopy method for TE (2) and TM (3) polarizations of light in the range of 600–700 nm.

In the case of the traditional, non-waveguide, excitation of the film using the illumination by the light beam with the intensity $I_1 = 10\,\text{mW/cm}^2$ in the direction normal to the film surface, the induced change of the absorption coefficient $\Delta\alpha \approx \Delta T/(Td)$ at 633 nm wavelength was about $10\,\text{cm}^{-1}$ ($k < 1 \times 10^{-5}$). From the measurements performed with the help of the waveguide technique, taking into account $\Delta k = k_2 \Delta I$ at $I = I_1$ and $k_2 = 1.15 \times 10^{-4}$ the absorption coefficient change is equal to 1.15×10^{-5}; here $\Delta I = I_1 - I_0$. As it follows from the data stated above, the measurement results of both the independent methods satisfactorily correlate. This allows one to use the proposed method for the measurement of the nonlinear constants of the Kerr media.

8.4. Optical Nonlinearity in Semiconductor Films

While fabricating the optical devices for the information processing of the materials, the refractive index and the absorption coefficient depending on the light intensity, are used. The high values of the nonlinear refractive index n_2 are typical for the semiconductor materials. Therefore, particular attention is given to the search and fabrication of semiconductor thin-film structures with high optical nonlinearity [253]. As the required controlling intensity at the low power of the incident light is easily obtained in optical waveguides due to their small cross sizes [259], so the optical nonlinearity in the guiding semiconductor structures is of definite interest.

8.4.1. Nonlinear Optical Properties of As_2S_3 Films

In the recent years, non-crystalline materials have a wide range of applications in different fields of engineering. The prospect of their use is determined to a great extent by the peculiarities of their optical and electrical parameters [308]. On the basis of the great volume of experimental data it has been assumed that chalcogenide semiconductor glasses (arsenic sulfide is considered here) are optically isotropic media. This fact assumes that structural elements forming these media are situated chaotically in the long-range ordering scale and this leads to the complete macroscopic isotropy of the properties of such glass [309]. Taking into account the remarks about the band structure of amorphous materials given in Chapter 1, let us consider the optical properties of chalcogenide semiconductor glasses (ChG). According to the experimental data the band gap for As_2S_3 is equal to $E_g = 2.32$–2.40 eV, for crystalline state $E_g = 2.70$ eV. The absorption coefficient up to 10^4 cm^{-1} is described quite accurately by the exponential Urbah law, when the value $\alpha < 10^3$ of its spectral dependence is described by the expression

$$\alpha = \alpha_1 \exp(-G\hbar\omega), \tag{8.4.1}$$

where $G = 18.6$ eV^{-1} and α_1 is a constant [60].

The refractive index of ChG at 630 nm wavelength is equal to 2.55. One of the peculiarities of the glass-like and amorphous semiconductors is their conduction of the hole type. The magnitude of this conduction changes relatively weakly at the doping of the semiconductor by most of the impurities, which affect the properties of the crystalline semiconductors strongly. But the great change in the conduction magnitude and also in the ratio between its electronic and hole components are caused by the influence of the surface state and its properties [310]. As it follows from the measurement results [60] the band structure of glassy As_2S_3 looks as in Figure 1.4. The Fermi level is fixed by the defects inside the band gap. Surface states can be localized and also so non-localized. The density of the states near the Fermi level quite often reaches values of 10^{16} cm^{-3} [238]. The width of this range is approximately 0.9 eV and the Fermi level position relative to the top of the valence band $E_F = 1$ eV (from 0.95 to 1.2 eV according to the results stated by various authors) [60]. The high density of the localized states is typical for chalcogenide glass semiconductors and together with their optical homogeneity this creates conditions for the use of these materials as nonlinear optical media [311]. Thus, the information about their nonlinear optical properties is a requirement.

Let us consider in more detail the way nonlinear optical constants are determined from the example of arsenic sulfide films on the substrate made from K8 glass, the films being obtained by the thermal evaporation in vacuum at gas pressure 10^{-5} Torr and at room temperature. It is reasonable to use single-mode He–Ne laser as the radiation source. We will excite the waveguide by the Gaussian beam (beamwidth $a = 70.4\,\mu m$), focused on the base of the measuring prism made of optical glass, e.g. TF5. Note that the spatial distribution of the beam intensity is not necessarily the Gaussian distribution. But in this case one needs to know analytically the function describing the intensity distribution. As the laser radiation has usually the Gaussian distribution in practice (see Section 3.2.) so it is natural to use this distribution in the process of determinating thin-film parameters. However, it is still necessary to control thoroughly the beam shape and its intensity distribution during the measurement. The power of the beam, which is incident onto the prism coupler (see Figure 8.8), is modified with the help of the attenuator (3) and controlled by the photodetector (11). The intensity distribution is recorded by the photodetector (12), placed in the focal plane of the lens (6). The determination of the intensity distribution extrema (see Section 3.3) assumes interpolation of the experimental data array by the function $b(\varphi)$ and its further investigation is based on the normal regression analysis [167].

To determine the waveguide linear parameters and the conditions of its excitation it is necessary to reduce the light power to the minimum that is determined by the sensitivity threshold of the recording device. Thereby, the power of the incident radiation in the considered example was reduced to the value of $0.5\,\mu W$. Then the beam axis was matched with the minimum in the distribution $b(\varphi)$, the distribution $b(\varphi)$ being usually a little asymmetrical because of the absence of the light intensity threshold for the nonlinearity of the Kerr type. To determine the linear parameters of the waveguide after the measurement of $b(\varphi)$ for several modes one has to use the averaging of the extrema in the $b(\varphi)$ distribution for every mode (Figure 8.12, curve 1) or the steepest descent method of optimization [168], which has been applied for the solution of inverse problem of the determination of the parameters of the guided mode guided mode. The processing of the series of the recorded Fourier spectra allows one to construct the dependence $\varphi_{\min}(I/I_o)$ that is linear with high accuracy in this case (see Figure 8.7, for example). By using the straight-line interpolation of this dependence one can

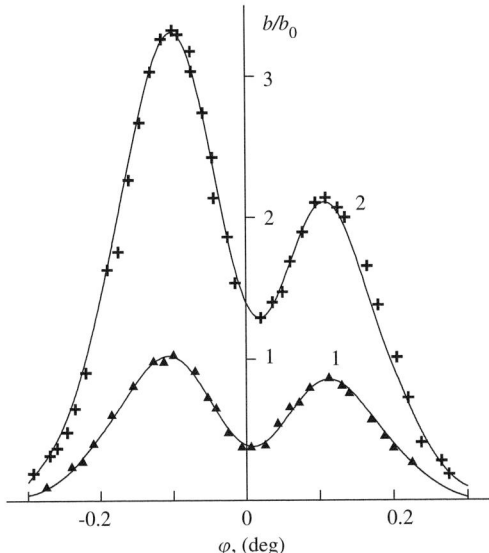

Fig. 8.12. Changes in the angular Fourier spectrum of the reflected beam at the consecutive increase in light intensity ($I_1 < I_2$); experimental data array is shown as points, distribution calculated on the basis of the model is shown as continuous curve [299].

get the value of φ_{min} at $I = 0$. The $b(\varphi)$ distribution corresponding to the given value of I_0 can be obtained when the beam axis is rotated out an angle $\Delta\varphi = -0.006°$, which is determined in the above described way. Having determined the propagation constants for any two modes, in this case, $h_7/k_o = 1.73002$ and $h_g/k_o = 1.54825$, one can find the thickness of the waveguiding film and its refractive index. For the arsenic sulfide film $d = 1.50\,\mu m$ and $n = 1.85073$. By using the series of the measured angular Fourier spectra for the reflected light beam it is possible to construct the $\varphi_0(I/I_0)$ dependence, and also the b_{min}/b_0 dependence as the function of I/I_0. The example of such dependencies is given in Figure 8.8. The processing of these dependencies allows one to determine the complex parameter of the nonlinearity of the waveguide structure.

After the calculation of the guided mode field, taking into account the determined nonlinearity parameter of the guiding structure $P_3 = 1.46 - i0.74$ we can find the nonlinear constants of the film. For films from As_2S_3 the nonlinear optical constants were $n_2 = 2.65 \times 10^{-3}\,cm^2/W$, $k_2 = 1.3 \times 10^{-3}\,cm^2/W$ [312].

For checking the feasibility of the obtained results one needs to evaluate the value of the variation of the film refractive index caused by the film heating by the absorption of the light beam energy. The example of the possible approach, for the solution of such problems, is given in Ref. [313]. The estimation of the maximum shift $b(\varphi_{min})$ due to the heating of the arsenic sulfide film, reveals the fact that this shift is five orders less than the experimentally observed value [312]. Thus, the influence of thermal nonlinearity is negligible in the example considered here.

By, using a single beam in the process of measurement it is possible to determine both the nonlinear optical constants of the thin film. But while comparising the obtained value of the nonlinear optical constant n_2 with the known data for chalcogenide glasses, n_2 appears to be significantly greater than the value mentioned in Ref. [314], for example. This discrepancy can be explained by the difference in light intensity ranges applied in the experiment. The light intensity value averaged over the cross-section of the waveguide in the range from 0.02 to 0.07 W/cm^2, while in Ref. [314] it was equal to $\sim 10^2$ W/cm^2. In the light intensity range $25 < I < 100$ W/cm^2 the $\varphi_{min}(I/I_o)$ dependence becomes less pronounced. In this case the n_2 nonlinear constant is equal to 1.5×10^{-5} cm^2/W and satisfactorily correlates with the results given in other references. The $h'_n(I)$ dependence in the range of all applied intensities is represented by curve 1 in Figure 8.13 [299].

The quantity h'_n used in this chapter is related to the measured resonant angle ϑ of the excited guided mode by the expression $h'_n = k_o n_p \cos \vartheta$. Introduction of some h'_n parameter, which is conventional in this case, is caused by some difficulties of interpretation of the propagation constant of the nonlinear guided mode in "prism–waveguide" structure. But in some works, e.g. in Ref. [315], such terminologies has been used. In the linear case the h'_n value is equal to the real part of the propagation constant of the

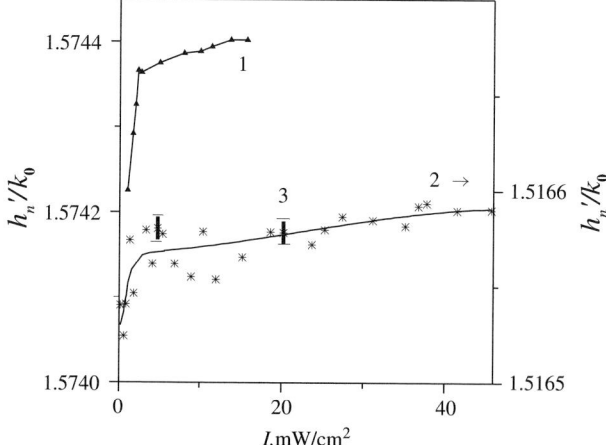

Fig. 8.13. $h'(I)$ dependence for As$_2$S$_3$ film (1) and ZnSe (2), (3) is the measurement error of h'.

guided mode. The corresponding dependencies of the output power and the reflection coefficient of the light beam on input power density are given in Figure 8.14.

The intensity range of the beginning of curve 1 in Figure 8.13 corresponds to the intensity range of curve 1 in Figure 8.14b. Considered peculiarities of the optical properties of the As_2S_3 film point to the existence of several optical nonlinearity mechanisms with different saturation intensities. Chalcogenide glassy semiconductors are materials with photo induced phenomena

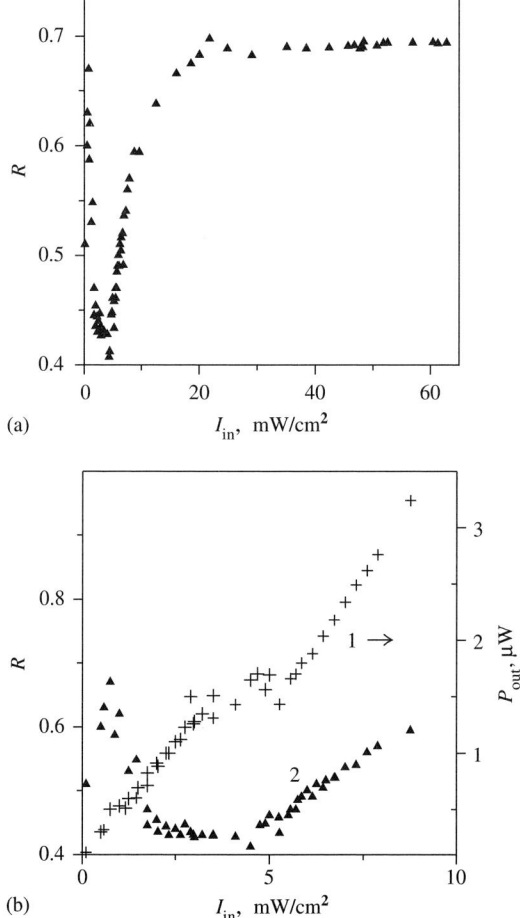

Fig. 8.14. Dependence of the light beam reflection coefficient (a) and output intensity on input intensity (b).

accompanied by photostructural transformations [316]. The photoinduced anisotropy, as observed in film samples [317], and also in monolithic ones [318], is typical for these materials. Nonlinear changes in the refractive index and the absorption coefficient of As_2S_3 films are observed at the excitation of the guided modes in the film by light with energy $\hbar\omega > E_g$, so $\hbar\omega < E_g$ [316,319].

It is reasonable to assume that the mechanism of nonlinear changes in optical properties in the case of As_2S_3 is caused by the interaction of the light with the electron states, which is related to the defects of the semiconductor, with energy levels situated in the band gap. Photoinduced changes of the optical properties in such materials can be explained by the fact that the excited non-equilibrium charge carriers are captured by the localized states inside the band gap. Thereby, the optical transitions from these states to the allowed band under the influence of the light photons with energy less than the band gap width [311] become possible, the variation in the absorption coefficient value being proportional to the total concentration of the occupied localized states [320]. The relaxation of the induced changes occurs mostly because of the captured carrier thermalization, which obeys the exponential "tail" of the localized states by means of iteratively repeated acts of carrier ejection into the band and following carrier capture by the traps [311].

The nonlinear optical effect taking place in As_2S_3 is similar in its origin to the optical nonliearity in the polycrystalline films ZnS [321,322]. There is an assumption about these ZnS polycrystalline films that the surface states of the crystallites of small size significantly contribute to the change in the nonlinear optical properties of thin films. Let us pay attention to this mechanism as one of those responsible for optical nolinearity in semiconductor films, and then consider the nonlinear optical properties of polycrystalline films.

8.4.2. Structure and nonlinear properties of ZnSe films

Studying the thin-film properties of zinc selenide (ZnSe) is of great interest because of the wide range of its possible applications as a material for effective nonlinear transformation of optical signals in data-processing devices [323–326]. Optical nonlinearity in polycrystalline semiconductor films manifests itself at the excitation of these films in the range of strong absorption and also when at the photon energy is smaller than its band gap [325]. As we are concerned with the nonlinear properties of the polycrystalline film depending on their structural characteristics [325], the study of the optical nonlinearity that depends on the crystal structure and quality of the deposited films is of definite interest. Some investigations of the optical properties and structure of zinc selenide films, obtained by RF sputtering of the ZnSe ceramic target, were performed from this viewpoint. Films with thickness up to 1 µm were deposited on substrates made of optical glass K8, quartz glass and sapphire (Al_2O_3 plane (0 0 0 1)) in the argon atmosphere, the pressure range 0.01–0.03 Pa and at the substrate temperature changing in the range from 350 to 550 K. The deposition rate was calculated by the measuring the film thickness. While studying the optical properties of the zinc selenide film the significant nonlinear changes caused by the variations of the incident light intensity were observed [327]. The dependence of the guided mode propagation constant on the incident light intensity is shown in Figure 8.13 (curve 2), where $\Delta h' = h'_n - h'$, with being h' the mode propagation constant at $I = I_0$, and I_0 the light intensity, the nonlinear effects of which are not being recorded yet. The significant dispersion of h'_n values that exceeds the measurement error ($\delta h' = 5 \times 10^{-6}$) for zinc selenide films requires more thorough investigation of the dependence of the parameters on the light intensity. The results of this study are depicted in Figure 8.15. It is obvious from the curves that the obtained dependencies have a complicated structure and are non-monotone in character.

The analysis of the optical and constitutive properties of films indicates the fact that the behavior of the non-monotone dependence and the value of n_2 of the deposited material are defined by the crystal structure of the deposited film. The temperature and the rate of the deposition are the critical

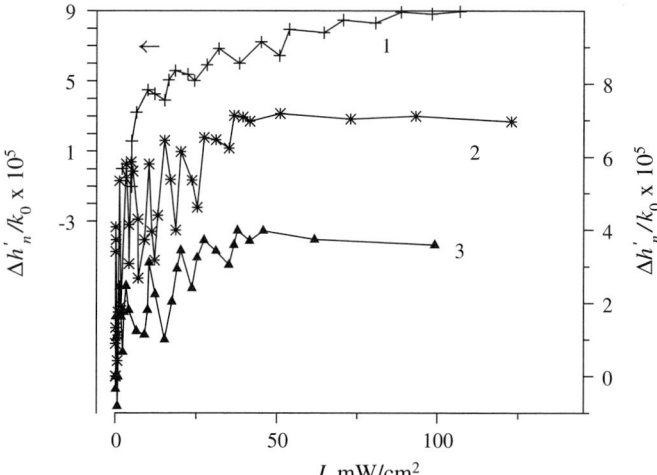

Fig. 8.15. Dependence of the h' change on light intensity for ZnSe films deposited at the substrate temperature 180°C (1), 240°C (2) and 280°C (3).

parameters that define the film properties for the prescribed kind of the substrate.

The dependence of the deposition rate on the substrate temperature is given in Figure 8.16 (curve 1, the substrate is made of quartz glass). The given dependence is typical for A_2B_6 film growth in the quasi-closed volume [328]. The decrease in the film deposition rate in the range of the substrate temperature up to 460 K can be explained by the increase in the desorption flux from the substrate surface. The increase in the deposition rate at the substrate temperature higher than $T_c = 460$ K is concerned with the increase in the surface diffusion coefficient of the deposited component, which, in turn, leads to the increase in the rate of the film deposition [329]. Thus, the minimum in this dependence is caused by the action of two opposite processes: the increase in the surface diffusion and the activation of desorption processes when the substrate temperature is increased. These processes define the structure of the deposited film. The results of the investigation of the film crystal structure with the help of X-ray diffractometer at the wavelength of monochromatic radiation of Cu–K_a are depicted in Figure 8.17. The typical diffraction spectrum of the ZnSe film is shown in Fig. 8.17a. Diffractograms of films deposited at various temperatures of the substrate are depicted in Fig. 8.17b.

The analysis of the given diffractograms indicates the fact that films are polycrystalline and in all the cases the crystallites have cubic structure with preferred orientation along (0 2 2) direction, which is parallel to the substrate. During identification of the diffractograms other structures were not observed. The additional information about the peculiarities of the film properties can be obtained from the measurements of the spectral dependence of the absorption coefficient. The results obtained by the method of

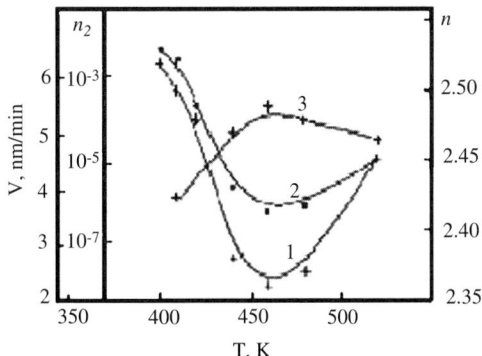

Fig. 8.16. Dependence of deposition of deposition rate (1), refractive index (2) and nonlinear optical constant n_2 (3) on substrate temperature.

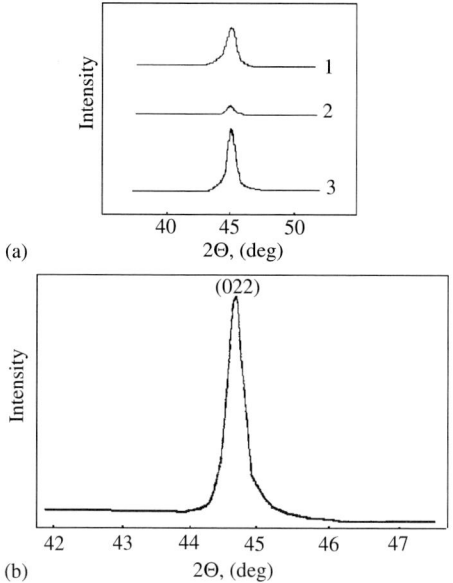

Fig. 8.17. Diffractogram of films deposited on the substrate made from quartz glass at various temperatures: (a) at 440 K; (b) 410 (1), 460 (2), 520 K (3).

traditional transition spectrometry with the help of spectrophotometer in the wavelength range 300–1000 nm for films deposited at various temperatures of the substrate, taking into account the reflection and interference effects, are given in Figure 8.18. The significant change in absorption near the wavelength of 450 nm, which can be considered as the edge of the optical band gap, is typical for all films. The presence of "steps" in the dependence of the absorption on the wavelength may be concerned with the dimension effects, which are caused by polycrystalline film structures [325]. The dependence of the film refractive index n, determined by the method described in Chapter 4 at wavelength of 633 nm, on the temperature of the substrate is presented in Figure 8.16 (curve 2).

The absorption coefficient for different films varied insignificantly and was equal to 2×10^{-4}–1×10^{-3}. While determining the nonlinear refractive index n_2 and the nonlinear absorption coefficient k_2 the results of the processing of the changes in the Fourier spectrum of the light beam reflected from the film in self-effect conditions at the variations of the light intensity at wavelength of 633 nm are used. The dependence of the nonlinear constant n_2 of the films on the temperature of its deposition is given in Figure 8.16 (curve 3). The nonlinear absorption coefficient k_2 has been changed in the

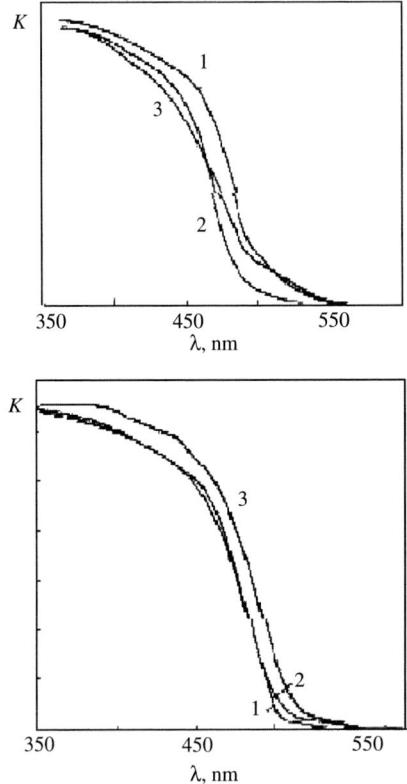

Fig. 8.18. Absorption spectra of films deposited on the substrate made from quartz glass (a) and on different substrates (b) at various temperatures. (a) 410 (1), 460 (2), 520 K (3); (b) quartz glass (1), optical glass K8 (2), sapphire (3); film thickness about 0.25 μm, $T = 440$ K.

range from 1×10^{-5} to 6×10^{-7} (the light-induced film blooming is observed) and, as it can be seen from the picture, it was also greater for films obtained at $T = 460$ K. The probe radiation intensity did not exceed the value of $10\,\text{W}/\text{cm}^2$ ($\lambda = 633$ nm). Thermal nonlinearity evaluated according to Ref. [312] was negligible. The performed investigations revealed the fact that at the deposition rate > 5.0 nm/min the films have low refractive index and the fuzzy edge of the optical band gap in the absorption spectra is shifted the long-wavelength range. This is, probably, concerned with the imperfection of the film structure and high concentration of defects, which are caused by the stoihiometry restrictions of the deposited material during the increase of the growth rate. At the deposition rate < 5.0 nm/min, the film's refractive index was close to that of the ZnSe monocrystalline in the wavelength range around 630 nm (2.52–2.58 [330]).

The increase in absorption near 450 nm is observed in the absorption spectra. The X-ray diffraction (XRD) data indicate the deposition of the oriented polycrystalline film of zinc selenide. At deposition rates <4 nm/min, the film parameters (the refractive index, optical band gap and crystal quality) are almost unchanged. Taking into account film quality and technological effectiveness of film deposition, the rate of 4.5–4.0 nm/min may be considered as optimal. Thus, all results stated below were obtained for films fabricated at this deposition rate. The presence of the minimum on the dependence of the refractive index on the substrate temperature (Figure 8.16, curve 2) and the observed shift of the absorption edge toward high frequencies in the absorption spectra for the films obtained at $T = 460$ K (Figure 8.18a) indicate the growth of fine-dispersed crystallites in films, the refractive index of the film material in this case being equal to 2.47. The wide peak at 44° in the XRD spectra, which is typical for zinc selenide and correspond to the orientation (0 2 2), is recorded in diffractograms (Figure 8.17b, curve 2).

Yedo et al. [331] observed a significant deterioration of the ZnSe film at the growth temperature <470 K. Properties of the films, deposited on zinc selenide substrates were also investigated. A significant improvement in the layer quality when the growth temperature decreased was also reported in this work. In this case improvement of film crystalline quality on foreign substrate at the substrate temperature <460 K was observed. The growth temperature of ~460 K may be called the critical temperature from the viewpoint of film quality. It is interesting that the films obtained at the growth temperature 410 K have better crystalline quality in comparison to films deposited at 470 K. This fact is illustrated by the XRD data (Figure 8.17). In the case of film deposition on substrates made from different materials the film properties differed significantly. For example, for the series of the substrate kind "sapphire–optical glass K8–quartz glass" the size of the lattice cell, evaluated on the basis of X-rays spectra, is decreased from 5.71 to 5.6 Å. The fundamental-absorption edge is shifted into the high-frequency spectral range (Fig. 8.17b) and the film refractive index is changed in the range 2.51–2.48 ($T = 440$ K). These differences are not so significant for the films with higher thickness. The reason for this is not quite clear. But we may agree with the Kalinkin et al. [329], who assumed that when the layer thickness exceeds some definite value there is a decrease in the substrate influence, and the growth of the films when thermodynamically stable cubic modification is to be expected. This should be noted, because it allows one to fabricate oriented films on different substrates at quite low temperatures.

While investigating the nonlinear optical properties it turned out that films, obtained under critical conditions, i.e. the films, deposited on substrates made from quartz glass at the substrate temperature near 460 K, have greater nonlinear constants. As nonlinear optical properties of polycrystalline films are often concerned with the dimension effects, caused by the influence of the crystallite boundaries [89], we can try to evaluate the influence of the crystallite size on the film properties. The evaluation may be executed in two ways, the results of which give will be in good agreement with each other. The crystallite size can be determined by the analysis of the XRD line broadening [332,333]. For zinc selenide films, this size has changed in the range from 19 to 7 nm at substrate temperature variations. The films, obtained at temperature 460 K, have the minimal crystallite size as can be seen from the analysis of the diffractograms. The crystallite size may be evaluated by the "blue" shift of the fundamental-absorption edge. These evaluations may be performed with the help of the effective mass approach [334] that takes into account the influence of the dispersion of the crystallite size [335]. The optical band gap is calculated by the absorption spectra on the assumption of straight transitions between the upper and the lower bands [43,150,336]. The application of this model, which is not quite adequate for synthesized structures, is not quite correct for the considered case. For this reason these calculations are of evaluative character. But the calculated crystallite size of 12–7 nm is close to the XRD data. And the films obtained at the critical temperature have also minimal crystallite size.

The decrease in the crystallite size leads to an increase in the surface state density in the films, and hence an increase in the nonlinear constant takes place. Zinc selenide is a diamond-like semiconductor and the chemical bound has a covalently ionic origin. The important peculiarity of this semiconductor is that it manifests only n-type conductivity [103,151]. As the percentage of ionic bounds in the semiconductor is quite significant, the defects behave as electrically active centers, in particular the vacancies in metalloid sublattice play the role of donors. It is postulated that in partially ionic crystals the band of acceptor type is given below the conduction band and the band of donor type above the valence band. The scheme of surface states, according to Ref. [58], is depicted in Figure 8.19.

In a general interpretation it should be mentioned that there is a band of filled acceptor levels near the valence band, but donor levels are usually free inside the conduction band. The thin-film structure properties define the position of the surface states. It should be noted that the surface state can be localized or resonant. The position of surface state levels above the valence band is $\simeq 0.67$ eV [58]. In this scheme the transition "surface state band" become possible at the absorption of light with the wavelength of 633 nm. Correlation of the complex nonlinear constant value and the crystallite size in deposited films allows one to consider the modification of the surface state system of the crystallite as the possible origin of the optical

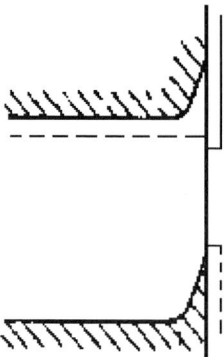

Fig. 8.19. Scheme of surface states in ZnSe.

nonlinearity in zinc selenide polycrystalline films when they are excited at a wavelength of 633 nm [338] by radiation.

8.5. Nonlinear Absorption in Semiconductor-Doped Glasses

The optical glasses doped by A_2B_6 semiconductors because of particular interest by the discovery of quantum-dimension effects and the influence of the interface on the optical properties of such materials [338,339]. The size of CdS_xSe_{1-x} crystallites distributed in the glass bulk controllably varies in the range from 1 to 100 nm. Optical nonlinearity in glasses was studied in the high-frequency range of the visible spectrum when photon energies $\hbar\omega > E_g$ [294], and when $\hbar\omega < E_g$ [340]. Under the influence of laser radiation there is a saturation of absorption in the spectral range from the absorption edge to the second quantizing level. The blooming is traditionally associated with dynamical reorganization of the crystallite energy spectrum at the excitation of non-equilibrium carriers [277,299,341]. Some authors note that the non-linear increase in the absorption in semiconductor nanocrystals is caused by the decrease in time of the non-radiative recombination of the carriers [342,343]. While investigating nonlinear effects in such media researchers often use OS12 glass, which is silica base doped by cadmium sulfoselenide compounds.

Optical properties of the films, obtained by RF sputtering of such glass under different deposition conditions, have also been studied in Ref. [344]. While investigating the optical nonlinearity by the waveguide methods in self-effect conditions the nonlinear dependence of h' on the intensity of the incident light was also observed in thin films (Figure 8.20.) The tendency of

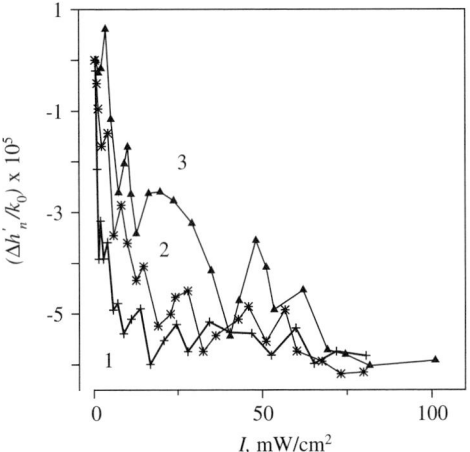

Fig. 8.20. Dependence of $\Delta h'(I)$ for films made from glass OS12 before (1) and after thermal annealing (3), deposited at the substrate temperature 140°C (1) and 190°C (2).

h' with the decrease and increase in light absorption by films remains. The optical nonlinearity in these materials is defined by the size of semiconductor crystallites incorporated into the glass matrix [251, 339]. The size of these crystallites can be changed by the thermal annealing or changing the conditions of film deposition [345–347]. The dependence of h' on the light intensity for films made from glass OS-14 before and after thermal annealing for 6 h at 400°C is depicted in Figure 8.20. The increase in the substrate temperature during the film deposition also modifies the behavior of the non-monotone dependence of the film's properties on the light intensity (Figure 8.20, curves 1 and 2). The absorption edge in the spectra of the studied samples is shifted after annealing in the long-wavelength range. This fact indicates the growth of the microcrystals in films [339,348] and of the decrease in the nonlinear constants of the film material. while the studying the properties of the films obtained from the low deposition rate, which provides the growth of the stoichiometric film, bleaching was observed. Note that the value of the nonlinear refractive index $|n_2|$ is decreased with the increase in the crystallite size.

It should be mentioned that optical nonlinearity when the intensity of light is low is also observed in the films made from the glass that does not contain impurities of cadmium sulfoselenide [344]. Thus, in the films obtained by RF sputtering of quartz glass the nonlinear changes in their optical properties are observed at the variation of the incident light intensity. The nonlinear refractive index is significantly less than the values of n_2 in semiconductor films or films made from semiconductor-doped glass, and equals $\sim 10^{-4}$–$10^{-7}\,\text{cm}^2/\text{W}$. But the nonlinear dependence of the optical parameters on the intensity of the incident light is reliably recorded only for films with broken stoichiometry. This effect is almost not observed at optical losses of the waveguiding film less than $\lesssim 2\,\text{dB/cm}$ (Figure 8.21).

The fact that nonlinear changes is parameters of the waveguiding film are recorded in the intensity range $< 10\,\text{mW/cm}^2$ follows from the data given in Figure 8.21.

It is rather difficult to explain such contradictory results. The study of the crystallites in the glass revealed that they have hexagonal as well as cubic structures [346]. The semiconductor impurities manifest the crysralline properties at sizes $>50\,\text{Å}$; otherwise these impurities manifest molecular properties [340]. At the same time the dimension effects appear when crystallite sizes are $<30\,\text{Å}$. A positive change in the refractive index in A_2B_6 hexagonal semiconductors is observed at the excitation by the low-intensity light beam in the range of the transparence of this semiconductor. This change is caused by the nonlinear polarizability of the bound electron. With further increase in the light intensity of the refractive index decreases due to the contribution of non-equilibrium carriers [349].

In cubic crystals of A_2B_6 type, only the refractive nonlinearity appears at the low excitation level. With the increase in the light intensity the two-photon absorption leading to the excitation of the non-equilibrium free carriers becomes apparent [350].

In this case the competition of different mechanisms of optical nonlinearity is possible. Such situation is typical for semiconductors [350]. The

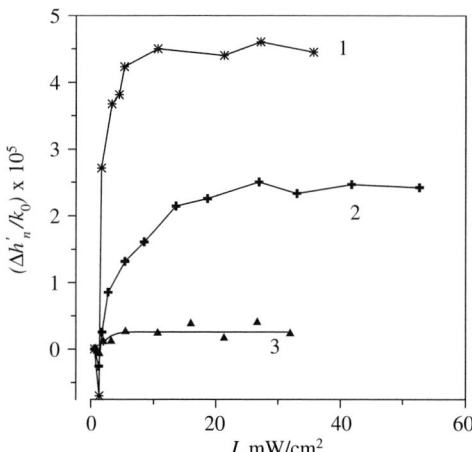

Fig. 8.21. Nonlinear changes of the waveguide properties of the films made from the quartz glass at the light intensity variations: $k = 2 \times 10^{-5}$ (1), 9×10^{-6} (2) and 3×10^{-6} (3).

analysis of the data presented above allows one to relate the observed effects with the modification of the surface state levels and also with the localized states in the band gap (Figure 8.22). To explain the mechanism of the absorption increase with the increase of the light intensity, some models were proposed. Such mechanism is observed during the illumination of the CdS_xSe_{1-x}-doped glasses. According to one of the models, new recombination centers appear in the semiconductor crystallites [351]. This becomes possible due to the migration of the photoinduced defect from the bulk to the surface of the crystallite. In another model the exit of the electron from the crystallite to the phase interface and its capture by the localized states in the band gap of the glass are considered [352–354]. The population of the localized levels in the glass host causes absorption in the range of the impurity absorption band. This is the reason for the appearance of additional absorption in the visible range of the spectrum.

At the same time the probability of the capture of the charge carrier by the crystallite defects is changed due to the presence of the occupied defect near the interface or as a result of the system electroneutrality change. Therefore, the results of the investigation of the glass host containing Cd, S and Se as ions but not as CdS_xSe_{1-x} crystallites are worth paying attention to. These works were done in order to obtain additional information about the mechanism, which is responsible for the photodarkening effect in the glasses with the nanocrystals [355]. The fact that the induced darkening can

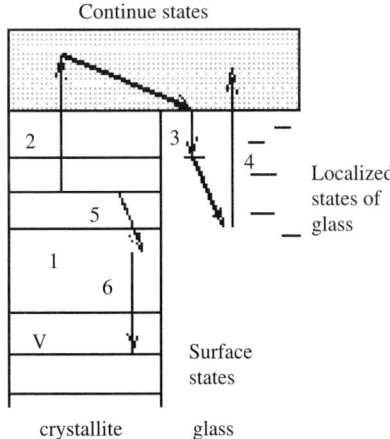

Fig. 8.22. Photo induced processes in the semiconductor-doped glasses [251]: (1) electron–hole pair recombination, (2) electron transition to higher level as a result of many-particle processes, (3) relaxation via the host states, (4) absorption via transitions between host states (photodarkening), (5, 6) recombination on defects or surface states.

be observed in the glasses without crystallites was also discovered. The only condition to be satisfied for this is the light wavelength to be in the range of the glass absorption. These results point to the second model that explains the changes in the properties of the semiconductor-doped glasses. The light absorption stimulates the generation of the charge carriers in the crystallite and intensifies the process of their capture by the defects that lead to the photodarkening. Otherwise, thermal processes cause inverse effect of the deionization leading to the restoration of the initial properties. But electron migration over the localized states in the host does not always end by the restoration of the initial states because of the stochastic nature of this process. Furthermore, in some works, for example Refs. [296,356] the decrease in the refractive index and the absorption coefficient was observed in the thin-film waveguides based on semiconductor-doped glasses. The negative value of the nonlinear constant n_2 can be explained by the mechanism of band filling. Banyai et al. [296] have observed non-monotone character of $n(I)$ dependence and found to depend on the saturation of the electron mechanism of the optical nonlinearity in these structures. They noted that thermal effects usually lead to an increase in the refractive index. The value of the nonlinear constant $|n_2|$ is decreases with the increase of crystallite sizes (Figure 8.21). It probably can be explained by the increase in the surface recombination area [357,358].

Thus, we can assume that surfaces and interfaces play a key role in the changes of optical properties of thin-film structures illumination by low-intensity radiation and with photon energy $\hbar\omega < E_g$.

Chapter 9
Optical Nonlinearity in Multilayer Structures

9.1. Features of Optical Properties of Multilayer Structures. 191
9.2. Interface Effects on the Nonlinear Optical Properties of Thin Films 197
9.3. New Potentialities for Studying Thin-Film Structures. 202

Optical nonlinearity in semiconductors is usually observed when the light power density equals at least 10–100 W/cm^2, for any mechanism, responsible for the existence of nonlinearity. In this chapter, we will discuss the observation of the changes in the refractive index and the absorption coefficient of thin films when the intensity of the incident light is <0.1 W/cm^2. The nonlinear optical constants measured reach values $\sim 10^{-3}$ cm^2/W.

9.1. Features of Optical Properties of Multilayer Structures

To create optical devices for signal processing or for optical communication, materials with large absorbing or dispersing nonlinearity are required. Low dimension and multilayer thin-film structures are of particular interest here. Owing to the periodicity of the spectral–optical parameters in the direction of the propagation of optical radiation the nonlinear dependence of properties of such structures on the incident light intensity is observed.

As a model system of current interest systems with multilayer structures, having a large number of interfaces, can be constructed. We will consider two kinds of multilayer structures: structures made of alternate deposition of lithium niobate (the conductor) and those made of quartz glass (dielectric). The conducting layers are isolated from each other by dielectric layers [362]. The number of layers in the multilayer structures varies from 2 to 20. The thickness of the films made from quartz glass and from lithium niobate is equal to 20–70 nm and 30–50 nm, respectively. The total thickness of the fabricated structures does not exceed 1 μm. The refractive index of the quartz film is equal to 1.476, the refractive index of films, obtained by RF sputtering of lithium niobate, is of 2.160 at the wavelength of 633 nm. While fabricating the multilayer structure, the thickness of the deposited layers can be controlled by the deposition time. But one should determine beforehand the film deposition rate at the sputtering of the material used for fabrication of the thin-film structure. We can estimate the accuracy of the determining a layer thickness by the deposition rate that varies from 0.2–0.3 to 10–30 nm/min, depending on the type of the deposited material. At the deposition of silicon dioxide films as the component of the multilayer structure, the targets of fused quartz are used, and in the case of the sputtering of lithium niobate the monocrystalline targets (z-section in this case) are used. In order to preserve the compound stoichiometry all films are usually sputtered in the argon and oxygen atmosphere (4:1). The substrate temperature does not exceed 180°C.

While determining the optical parameters of a thin-film structure (the refractive index, absorption coefficient and thickness), we can utilize the methods described above. It is obvious that the value of the refractive index is averaged over the mode profile, i.e. over some effective parameters. But the thickness of such thin-film structure determined by the waveguide method satisfactorily correlates to the mechanical stylus data ($\delta d = 100 \text{ Å}$) or interferometry ($\delta d = 50 \text{ Å}$). This indicates the accuracy of the application of the Fourier spectroscopy of guided modes for the investigation of these structures [381]. The nonlinear refractive index n_2 and the nonlinear absorption coefficient k_2 are measured in the self-effect case (no additional illumination) at the wavelength of 633 nm by the waveguide method also. The variations in the Fourier spectrum of the light beam reflected from the prism are the result of the gradual increase in the incident light intensity. The power of the incident radiation varies in the range from 0.5 to 500 µW. The beamwidth on the base of the prism coupler usually does not exceed 150 µm.

In the process of thin-film waveguide analysis the significant influence of the substrate parameters on the waveguide properties was revealed [359, 360]. For this reason the multilayer structures on the base of the semiconductor surrounded by the media with different optical parameters were fabricated. Let us consider the peculiarities of the properties of the two kinds of multilayer structures, depicted in Figure 9.1, where dielectric layers fabricated by RF sputtering of the target from the quartz glass and lithium niobate are indicated by 1 and 2, respectively, and the substrate made from quartz glass by 3.

We will examine the properties of structure 1 (SiO_2–semiconductor–SiO_2) and structure 2 (semiconductor–SiO_2–semiconductor). Structures of these kinds are fabricated under identical conditions during the one sputtering process. As an example, let us consider the modification of the optical properties during the variation of the intensity of the probe light for the structures obtained by the deposition of lithium niobate and quartz glass containing 11 layers. The first structure consists of five layers obtained by

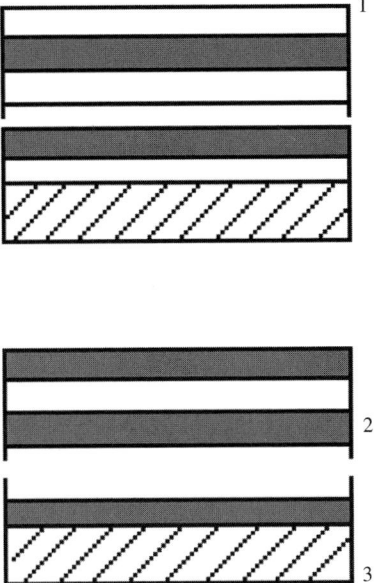

Fig. 9.1. Multilayer structures fabricated by the alternate deposition of lithium niobate and quartz glass.

the sputtering of the lithium niobate target and six layers made from quartz glass; the second one consists of six lithium niobate layers and five quartz layers. In this case the highly refractive layer has interfaces with quartz glass, air and the SiO_x layer. The transition spectrum of this structure is depicted in Figure 9.2. These structures manifest pronounced nonlinear optical properties [360, 361]. Note that such structures were also studied by other authors, who pointed to their non-ordinal properties [362]. The dependence of waveguide properties on the intensity of the incident radiation for structures of both kinds is shown in Figure 9.3 [363]. The dependence has a nonmonotone behavior and the curve $h'(I)$ has 5 and 6 extremuma for structures containing 5 and 6 layers, respectively, obtained by the sputtering of the lithium niobate target. The significant dispersion of h' greatly exceeds the measurement error $\delta h' = 5 \times 10^{-6}$.

For the structure with three layers there are three spikes in the $h'(I)$ curve (Figure 9.3b, curve2). A similar behavior of the h' dependence can be observed in the case of the excitation of guided modes of different orders and different polarizations by using the results obtained from the processing of variations in the Fourier spectra of guided mode during the gradual increase in the light intensity the nonlinear refractive index n_2 and the nonlinear

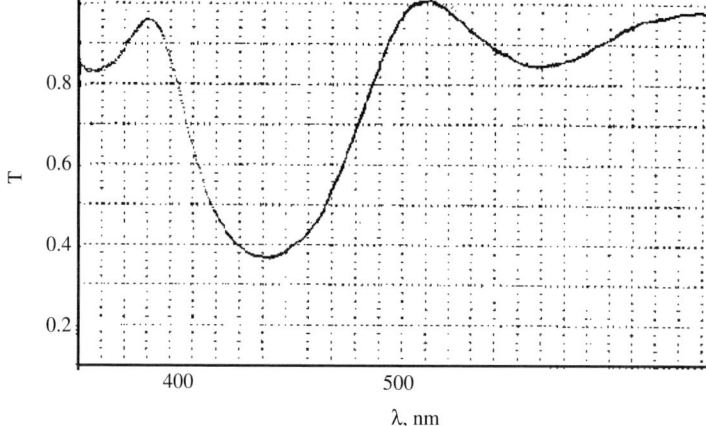

Fig. 9.2. Transition spectrum of multilayer structure on the base of lithium niobate and silicon dioxide.

absorption coefficient k_2 in different intensity ranges can be determined:

$$n_2^{(I)} = -2.1 \times 10^{-3} \text{ cm}^2/\text{W}, \quad k_2^{(I)} = -6.2 \times 10^{-3} \text{ cm}^2/\text{W},$$
$$n_2^{(II)} = 3.1 \times 10^{-3} \text{ cm}^2/\text{W}, \quad k_2^{(II)} = 5.1 \times 10^{-3}$$

(see Figure 9.3a, curve 2).

The dependence of the reflection coefficient on the intensity of the incident radiation has a non-monotone behavior as well as in this case (Figure 9.3a, curve 3). These results are for the TE mode of the first order. The light beamwidth is equal to 150 μm. Higher values of n_2 and k_2 at small light intensities allow one to use such structures as the nonlinear optical medium.

As lithium niobate is a nonlinear material, attempt was made to model the nonlinear optical medium by the fabrication of multilayer structure on the base of the linear optical materials in order to show reveal the effect of interface influence. These multilayer structures were fabricated by the alternate deposition of linear materials: conductive tin dioxide and dielectric SiO_2 [361]. The dependence of changes in h' on the changes in light intensity I for structures containing three tin dioxide layers separated by the quartz glass layers is depicted in Figure 9.4.

This structure with a layer thickness of about 10 nm is a good model of a low-dimension nonlinear medium. According to the stated results, the complex behavior of the dependence $h'_n(I)$ is determined by the number of layers in the studied sample. In the structure based on tin oxide the thickness of the conducting layers was equal to 12, 24 and 36 nm. Three spikes of different

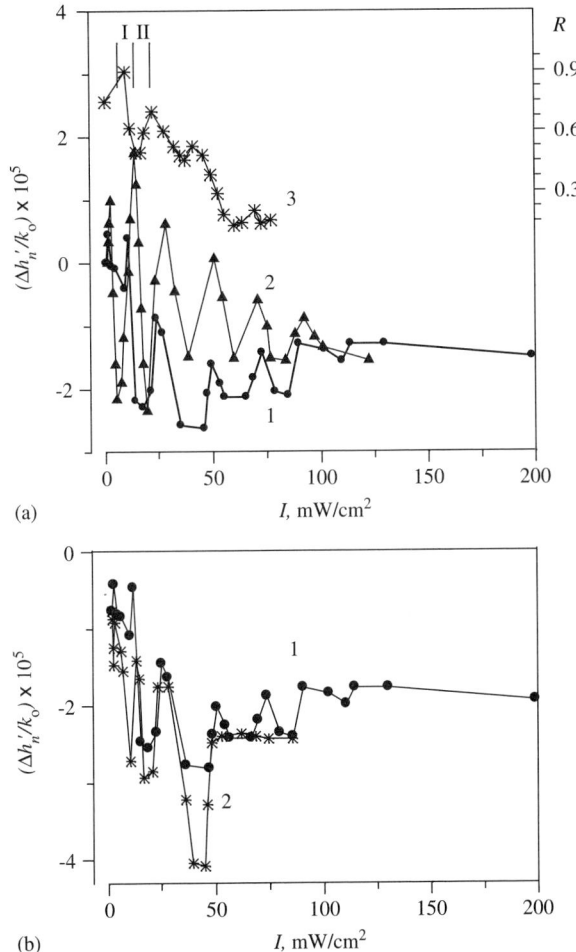

Fig. 9.3. Changes in h' at the incident light intensity variations for the mode of the first order in (a) multilayer structures of the first (1) and second (2) kind that contains lithium niobate and the light reflection coefficient (3) at the excitation of the first structure; and (b) multilayer structures of the first kind that contain six (1) and three (2) lithium niobate layers.

widths on the curve $\Delta h'(I)$ can be seen, i.e. the larger spike width of the curve $h'(I)$ corresponds to the larger layer thickness. The shape of $h'_n(I)$ dependence in the range of the third spike ($d_1 = 36\,\text{nm}$) looks like the analogous dependence for the film with larger thickness ($d_1 = 120\,\text{nm}$) depicted in Figure 9.5 (curve 3). The nonlinearity of the optical properties of the thin-film structure essentially depends on the optical quality of the dielectric

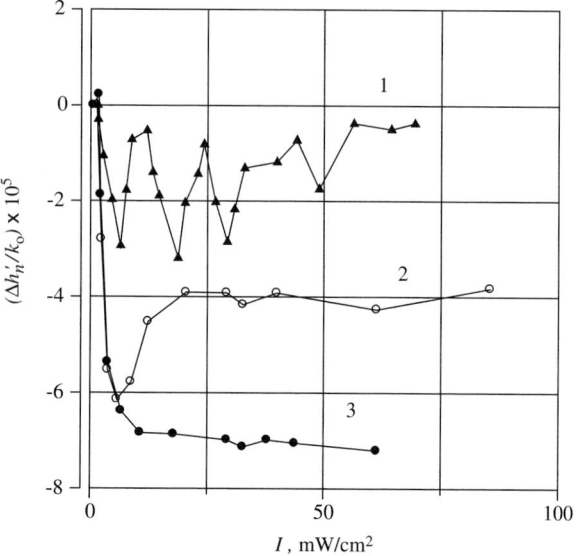

Fig. 9.4. Dependencies of h' variation for multilayer structures containing three conducting layers from tin dioxide (1), in air atmosphere (2) and in water vapor (3).

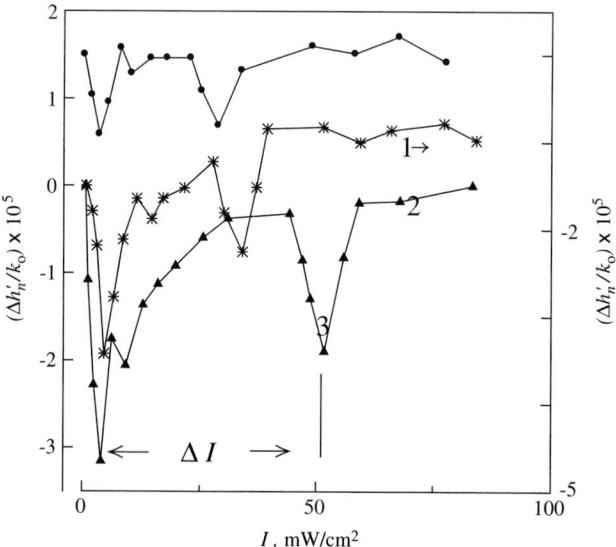

Fig. 9.5. Dependencies of $\Delta h'(I)$ for multilayer structure "SnO$_2$ substrate" (1), "SnO$_2$–SiO$_x$ substrate" with absorption coefficients of the SiO$_x$ film equal to 1.5×10^{-5} (2) and 5×10^{-6} (3).

layer. The dependencies of $h'(I)$ for three SnO_2 films deposited on different substrates during the deposition process are given in Figure 9.4 [361a], quartz glass and the structure "quartz glass–SiO_x film" being used as the substrate. Dielectric films are obtained under different deposition conditions and they have some differences in stoichiometry of the compound; therefore the absorption coefficient of the SiO_x film is equal to 1.5×10^{-5} (curve 2) and 5×10^{-6} (curve 3), and the SiO_x film thickness is equal to 1 µm.

The fact that even amorphous substrate affects the linear parameters of the waveguide film was reported previously [359]. In this case, the nonlinearity of the optical properties is brilliantly expressed in the waveguide structures, which contain imperfect SiO_x film with higher values of the absorption coefficient as the buffer layer between the substrate and the waveguide. Recall that optical nonlinearity was also observed in dielectric films at low-intensity radiation [344].

An analysis of the considered data allows one to assume that the region of the photoinduced absorption (see Figure 9.3a, range I) can be associated with the trapping of the light-generated charge carriers by the energetic levels of the localized states in the silicon dioxide film. These results indicate that the non-monotone behavior of the dependence of the optical properties of the multilayer structure on the intensity of the probe light beam is caused by the interfaces of "semiconductor–dielectric".

9.2. Interface Effects on the Nonlinear Optical Properties of Thin Films

An analysis of the dependence of the optical properties on the intensity the probe radiation was performed for glassy arsenic sulfide film, polycrystalline films of zinc selenide, zinc oxide, gallium selenide [371, 361a, 364] and the semiconductor-doped glass films [361a]. The fundamental absorption edge for these materials did not always coincide with the radiation wavelength equal to 0.633 µm, but the nonlinear variations of optical properties were observed in all the structures. Therefore, one of the possible reasons for the existence of optical nonlinearity is the surface and the interface effects [251]. The results given above indicate that it is the surface that causes the nonmonotone behavior of the dependence of the optical properties on the low intensity for such structures. The decrease in h' during the gradual increase in the light intensity is concerned with the decrease in the film material refractive index. The change of the absorption coefficient is also recorded in this case. These changes of thin-film parameters with the increase in the intensity of the incident light are often associated with the decrease in the concentration of the charge carrier in semiconductors [70].

It is possible to obtain additional information about the processes that cause the nonlinear dependence of the optical properties of thin films from the spectral measurements. For example, additional illumination of the studied sample by the second laser beam with wavelength $\lambda = 633$ nm changes the spectral absorption coefficient of the film material. The absorption spectra of the SnO_2 film before and during the additional illumination by the laser beam are shown in Figure 9.6. The spectra are measured by the waveguide spectroscopy method. The main principles of this method are described in Chapter 6. The absorption spectrum of the same film in water vapors is also depicted in Figure 9.6 (curve 3). An analysis of these results points out the existence of the occupied and the empty energy levels of the surface states in the spectral range of transparence of film materials. The data of photomodulation spectroscopy depicted in Figure 9.7 confirm this fact. The film spectra measured by the spectrophotometry method are also shown in the figure (curve 2).

Probably, the existence of the surface states determines the significant nonlinearity of the spectral–optical parameters of semiconductors and dielectric thin films in the range of the low intensity. If the non-monotone behavior of the $h(I)$ dependence is associated with the modification of electron states caused by the surface, it can be modified by the injection of gas

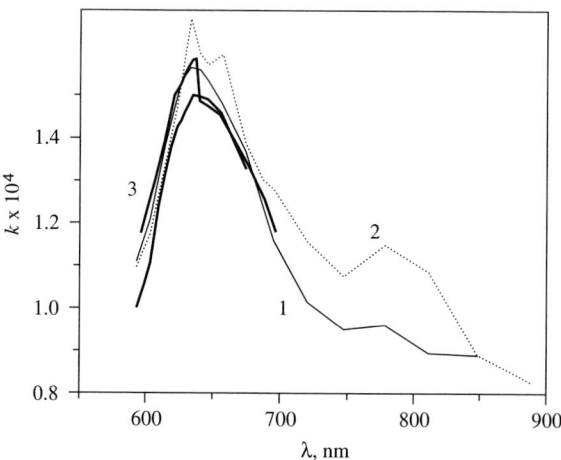

Fig. 9.6. Absorption spectrum of the SnO_2 film before (1) and during the additional illumination by the light at wavelength 632.8 nm (3), in water vapors (2), without additional illumination.

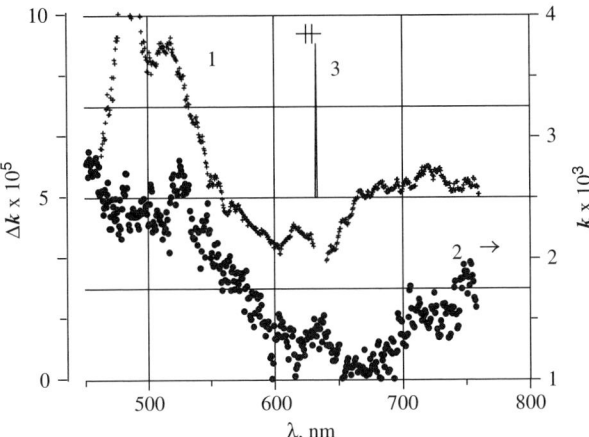

Fig. 9.7. Photomodulation absorption spectrum (1), absorption spectrum of SnO_2 film (2), the radiation modulated with the frequency 4 Hz (3) (according to Dr. A.I. Voitenkov).

impurities into the surrounding medium. Such an experiment can be realized under the conditions of total internal reflection in the waveguide, when the interference of the light beam incidental on the interface and reflected from it causes the standing wave in the media surrounding the waveguide.

The field of the mode penetrates into the medium surrounding the waveguide and interacts with the gas molecules. It is known that the actual area of the film's surface can exceed its geometrical area. This can be explained by the existence of an intrinsic surface in the film that creates vacancies in structure [365]. Gas molecules adsorbed on the surface of thin-film structures can either create additional states and cause annihilation of the energy levels of the surface states [41]. To perform such investigations the results given in Chapter 7, and the waveguiding structure is similar to the one depicted in Figure 7.5, can be used. This structure consists of silica dioxide and tin dioxide layers, which are consecutively deposited onto the base of the glass prism. In this case, the gas-impurity molecules can be added to the semiconductor–film surface. It is known that the water vapors actively deactivate the surface state levels [177]. So it is quite natural to assume that while performing such measurements in the atmosphere of water vapors the effect of the decrease in absorption in the film (and increase of h') will be depressed. The results of the investigations of the dependence of waveguide film parameters on the incident light intensity in the atmosphere of water vapors and in air are depicted in Figure 9.7 (curves 2 and 3). The range of curve 3, where the refractive index of the film is decreased and the absorption increased, remains unchanged at low intensities of the incident light, and the blooming is not observed when the intensity is further increased at the contact of the waveguiding film with water vapors.

Therefore, the assumption about the influence of the interface on the nonlinearity of optical parameters of a multilayer structure is valid. An analysis of the given experimental data and the results of other investigations [366, 367] allow one to assume that optical nonlinearity in such structures is caused by the electron processes at the "semiconductor–dielectric" interface and the appearance of nonlinearity is associated with the modification of the electron levels of the surface states. If the condition $|k_2^{(I)}| > |k_2^{(II)}|$ is satisfied for the structure (see Figure 9.7), the processes taking place at the interfaces of layers increase the total absorption in the multilayer structure with the increase in the incident light intensity. Otherwise, if $|k_2^{(I)}| < |k_2^{(II)}|$, the effect of film blooming during the growth of the incident radiation intensity is observed. It is evident from the analysis of the data stated above that the common tendencies for dependencies of the optical properties on the light intensity were observed in amorphous and polycrystalline semiconductor films, dielectric films, films made from semiconductor-doped glasses, and multilayer structures.

Let us consider this process in detail. With the illumination of the semiconductor the generation of electron–hole pairs and the diffusion of these pairs into semiconductor bulk take place. When the semiconductor film thickness is greater than SCR, charge carriers are generated at the surface, or rather charges are generated in the layer near the semiconductor surface. In this case the equation for the balance of charge carriers near the surface will be

$$g_s = \frac{1}{e} j_s(O) + S \delta n_s,$$

where g_s is the rate of the surface generation, $S \delta n_s$ the rate of the decrease in the charge carrier caused by the surface recombination, and $j_s(O)$ is the current of charge carriers near the surface [43].

The equation is written assuming that g_s does not depend on the absorption coefficient of the film material. Since the spectral frequency of the incident photons is situated in the transparence range of the films and the concentration of the surface states can reach values equal to the number of atoms at the film surface, $j_s(O) < 0$. In this case, the charge carrier flux is directed to the semiconductor surface. It partially blocks (i.e. saturate) the surface states, which causes the variation of the optical properties of the films. Since not all of the surface levels are involved in the recombination processes and only "band-level" transitions are accessible for part of states, some surface levels become traps. These states are more typical for the interface between the semiconductor and the dielectric film. In the area of heterojunction of the semiconductor and the dielectric film there are charged centers, which create the surface traps with cross-sections of small free-carrier captures [368, 369]. Even if there is a qualitative interface, as in the case of "SiO_2–Si" structure, the dielectric film always contains a significant incorporated charge that is located near the interface of the layer with the thickness <30 Å [238]. As a result, there is a potential barrier near the interface. This barrier localizes electrons and holes near the semiconductor surface. Traps can be either fast or slow [370]. The crowding of the localized states causes additional light absorption in the corresponding spectral range. At the same time the possibility of charge carrier trapping is changed because of the existence of "filled" defect near the interface [354]. Besides, note that not all of the surface states are localized and part of the states is involved in the recombination processes. The filling of the surface state leads to a decrease in the absorption in the film. Thereby, the process of photoinduced darkening in the thin-film structures can be explained by the example of a two-layer energy model of optical recharge of localized states, which are situated in the band gap [371], and the film "bleaching" can be explained by the band- "filling" effect [253]. Competition in the optical nonlinearity mechanisms occurs in the experiment due to non-monotone dependence of the film parameters on the light intensity [350]. The presence of these processes probably causes the non-monotone character of the dependence of the optical properties of the structure in the range of low intensity of the incident light used for the excitation of the structure in the range of transparence [351].

The non-monotone dependence of the reflection coefficient on the intensity in case of the reflection from the interfaces of absorbing film is theoretically investigated in Ref. [372]. Generally, the dependence of the absorption coefficient of the thin-film structure on the intensity of the incident light is defined by the correlation between rates of recombination and other relaxation processes [261]. The data stated above show the similarity of the processes taking place at the surface of semiconductors and dielectric films, in multilayer structures and films made from semiconductor-doped glasses. These processes explain the non-monotone dependence of optical properties of such films on radiation intensity. The nonlinear properties of the investigated structures are caused by the drift of the charge carriers generated by the light from the semiconductor bulk to the interface, by their trapping on the energy levels inside the band gap, and by the recombination processes of non-equilibrium carriers also. The presence of these competitive processes at the interface leads to the non-monotone dependence optical properties of thin-film structures on the intensity of the probe light beam.

The results stated in this section allow one to explain the nonlinear variations of the optical properties of the thin-film structure at the photons below the band-gap energy as a result of the process of the excitation, trapping and recombination of carriers at the interface. The conclusion that it is the behavior of the interface in thin-film structures that determines the nonlinear dependence of their optical parameters creates new perspectives in the investigation of low-dimension structures.

9.3. New Potentialities for Studying Thin-Film Structures

The results stated in the previous chapters show that the methods of investigation of linear and nonlinear properties of thin films allow one to obtain the information that quite well correlates with data obtained by other techniques. In conclusion, we will consider the application of the Fourier spectroscopy of the guided modes for studying low-dimension structures.

During the investigation of waveguiding thin-film structures a giant optical nonlinearity of Kerr type ($n_2 \sim 10^{-3}$ cm^2/W) was revealed at the light intensity < 0.1 W/cm^2. The dependence of the waveguiding and optical properties of thin films on the light intensity had a non-monotone behavior. Therefore, while investigating the nonlinear properties of quite thick glassy arsenic sulfide film (thickness from 1.5 to 5.0 µm) the non-monotone dependence $h'(I)$ was obtained. The dependence of the light reflection coefficient and the intensity of the reflected beam at the excitation of the guided mode with the help of the prism coupler had also a non-monotone behavior with the gradual increase in the light intensity (see Figure 8.14). The analysis of the results obtained for zinc selenide polycrystalline films shows that nonmonotone dependence (see Figure 8.15) and the value of n_2 for deposited films depend on the crystalline quality of the thin film. The examined films were polycrystalline and the crystallites had cubic structures with the preferred orientation along (022) direction parallel to the substrate. As follows from the XRD data, the studied polycrystalline films had average sizes of separate crystallites equal to 7–12 nm. Similar dependencies were obtained for low-dimension structures fabricated from the semiconductor-doped glasses (see Section 8.5). The origin of the recorded optical nonlinearity is defined as the photoinduced modification of the surface-state system. In this case an attempt to model the investigated object as multilayer structure that contains large number of interfaces was made. The dependence of h' on the incident intensity for these structures was also non-monotone in character. The complex behavior of the $h'(I)$ dependencies is defined by the number of layers in the investigated sample [358, 362], even if these layers were obtained by the deposition of the same material. For example, the SnO$_2$ film was obtained in two steps, i.e. two layers were deposited under identical conditions with a delay of 2.5 h, and the sample was not affected by the atmosphere. The break was vividly seen in the $h'(I)$ dependence and is concerned with the interface between these layers. The fact that the wider spike on the $h'(I)$ curve corresponds to the greater thickness of the layer (see Figure 9.4) was discovered. At the thickness of the nanolayer of more than 30 nm the nonlinear response from the surface of each separate layer was recorded. One more fact should be mentioned here: the optical quality of the waveguiding film affects the behavior of these dependencies (e.g. see dependence of optical properties of the quartz film on the intensity of the incident light in Figure 8.21).

The photoinduced absorption in the semiconductors is usually associated with the trapping of electrons on the localized states inside the band gap. The origin of these localized states is often concerned with the defects at the interface or film surface. For this reason it is possible to measure the nanolayer thickness in the low-dimension structures and to perform the investigations of the spatial distribution of the defects near the interface in the thin-film structures. The results of the processing of the dependencies mentioned in this chapter are depicted in Table 9.1.

Table 9.1. Parameters of dependence $h(I)$ and thin-film properties

Thin-film structure	ΔI (mW/cm^2)	d_1^* (nm)	n_2 (cm^2/W)	d_1 (nm)
SnO$_2$ film				
$k_0^{-1}h''$				
3.14×10^{-4}	229	707	1.9×10^{-3}	693
2.1×10^{-4}	26	117	1.8×10^{-3}	111
2.4×10^{-4}	30	117	1.9×10^{-3}	118
3.9×10^{-4}	45	117	2.0×10^{-3}	115
SnO$_2$ multilayer structure,				
$k_0^{-1}h'' = 7 \times 10^{-4}$	8.8	12	2×10^{-3}	12.5
	17.1	24		24.4
	25.4	36		36.3
SiO$_x$ film				
$k_0^{-1}h'' = 2.9 \times 10^{-6}$	22	1050	2.8×10^{-4}	1062
LiNbO$_3$ multilayer structure,				
$k_0^{-1}h'' = 2 \times 10^{-4}$	10	50	2.1×10^{-3}	52.5
As$_2$S$_3$ film,				
$k_0^{-1}h'' = 8.43 \times 10^{-5}$	100	1500	2.6×10^{-3}	1542
Polycrystalline film ZnSe,				
$k_0^{-1}h''$				
1.4×10^{-4}	2.6	7	8×10^{-4}	7.4
1.2×10^{-4}	5.0	12	6×10^{-4}	12.5
9.0×10^{-5}	7.4	19	4.5×10^{-4}	18.5

*The values obtained by other methods.

In the analysis of these data, one can see that there is a vivid correlation between the values of spike width ΔI (see Figure 9.4) on $h(I)$ dependence curves and a certain layer thickness d_1. The nanolayer thickness can be evaluated from the following expression [373]:

$$d_1 = C|n_2|\Delta I/2h'' \tag{9.3.1}$$

where $C = 9.92$ is a constant.

The phenomena being observed can be explained considering the excitation of guided modes in the structure with the help of a prism coupler. Obviously, the excited mode propagates in all layers of the structure and the mode field distributions for various light intensities are identical (at least, at small values of I). But when $I = I_1$ (see Figure 9.8) the mode amplitude is so small that the nonlinear effects have not yet apparent, or more correctly, are not recorded by the available setup.

With the gradual increase in the intensity the nonlinear response from the first interface, then from the second, and soon can be recorded. The experimental data revealed that the nonlinear response from the "near to the prism" interface is usually recorded first. When the light intensity exceeds a certain value, the saturation of electronic mechanisms of the optical nonlinearity in the near-surface layer of the structure occurs and the propagation constant becomes independent of the light intensity. Thus, it is possible to speak about an analogy with the tunnel microscope, whose "depth of focus" varies with the increase in the incident light intensity. These speculations can be applied to the polycrystalline films also. But in this case the semiconductor crystallite surface should be considered as an

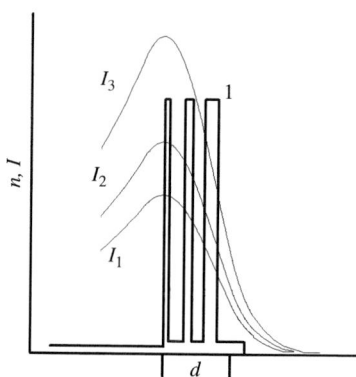

Fig. 9.8. Mode field distribution for three-layer structure on the basis of SnO_2 and the refractive index distribution over the structure depth (1).

interface. By analogy with the multilayer structures, one can assume that the greater width of the extremum in the $h'_n(I)$ function corresponds to the greater crystallite size (cf. compares Figures 9.4 and 8.19). Here, the crystallite size is the size in the direction, which is normal to the film surface, and this size is averaged over the light beam cross-section.

It is to be noted that the measurement capabilities of the standard methods employed for investigation of such structures, e.g. electron microscopy or X-ray diffraction, are restricted. The smallest and often the most interesting objects cannot be observed because of their small size and restricted contrast of the recorded signal from the film semiconductor crystallite [374, 375]. At small-angle scattering the intensity of the reflected radiation is proportional to r^6, where r is the radius of the particle being observed. Therefore, the intensity of the radiation scattered on the density oscillations in the film can mask the scattering caused by the crystallite due to its small size [376]. The resonant methods may be useful for studying the lowered dimension structures [377]. Besides, the joint analysis of the results obtained by different methods can be used in order to avoid ambiguities that appear at the interpretation of separate measurements.

Thus, the methods of the waveguide spectroscopy stated in this book give the possibility to perform complex investigations of low-dimension thin-film structures. With the use of simple equipments they allow one to evaluate the nanolayer thickness or crystallite size in such structures.

References

1. V. V. Ioffe, "Refractometry Methods in Chemical". Chemistry, St.-Peterburg, 1983.
2. A. S. Vasil'ev, "Metrology and Technical Measurements". Mashinostroenie, Moscow, 1988.
3. M. A. Bukhstab, "Measurements of Low Losses". Energoatomizdat, St.-Peterburg, 1988.
4. N. P. Gvosdeva, K. I. Korkina, "Theory of Optical System and Optical Measurement". Mashinostroenie, Moscow, 1981.
5. R. Azam, N. Bashara, "Ellipsometry and Polarized Light". Mir, Moscow, 1981.
6. G. J. Dizon, *Laser Focus World* 7, 107 (1998).
7. A. I. Belyaeva, A. A. Galusa, T. G. Grebennik, *Opt. Spectrosc.* 87(6), 1041 (1999).
8. J. Bartella, J. Schroeder, K. Witting, *Appl. Surf. Sci.* 179, 181 (2001). V. A. Shvec, S. V. Rykhlitsky, *Avtometry* 1, 5 (1977).
9. D. Zankovsky, *Laser Focus World* 1, 111 (1999).
10. A. N. Basil'ev, V. V. Mikhaylin, "Introduction in Dielectric Spectroscopy". Yanus-K, Moscow, 2000.
11. V. V. Philippov, *Opt. Spectrosc.* 88(4), 641 (2000).
12. M. M. Gurevich, "Photometry: Theory, Techniques and Devices". Energoatomizdat, St.-Peterburg, 1983.
13. N. Haryk, "Spectroscopy of Internal Reflection". Mir, Moscow, 1970.
14. A. C. Boccara, D. Fournier, J. Badoz, *Appl. Phys. Lett.* 12, 72 (1980).
15. A. A. Betin, O. V. Mitropol'cky, V. P. Novikov, M. A. Novikov, *Sov. J. Quantum Electron.* 12(9), 1856 (1985).
16. P. A. Gass, S. Schalk, J. R. Sambles, *Appl. Opt.* 33(31), 7501 (1994).
17. D. Klijer, Ed., "Super-sensitivity Laser Spectroscopy". Mir, Moscow, 1986.
18. V. S. Burakov, V. V. Zhukovsky, A. A. Stavrov, *Sov. J. Quantum Electron.* 5(1), 13 (1978).
19. V. M. Baev, V. P. Dubov, E. A. Sviridenkov, *Sov. J. Quantum Electron.* 12(12), 2490 (1985).
20. T. A. Golubkina, A. A. Tischenko, *Int. Radioelectr.* 9, 37 (1987).
21. E. M. Zolotov, V. A. Kiselev, V. A. Sychugov, *Sov. Phys.-Uspekhi* 112(2), 231 (1974).
22. K. A. Landa, G. T. Petrovsky, "Amorphous Planar Waveguides". Krasn. State Univ., Krasnoyarsk, 1987.
23. J. E. Midwinter, *IEEE J. Quantum Electron.* QE-7, 339 (1971).
24. P. R. Tien, R. Ulrich, R. J. Martin, *Appl. Phys. Lett.* 14(9), 291 (1969).
25. R. Ulrich, R. Torge, *Appl. Opt.* 12(12), 2901 (1973).
26. R. T. Kersten, *Opt. Acta* 22(6), 503 (1975).
27. J. S. Wei, W. D. Westwood, *Appl. Phys. Lett.* 32(12), 819 (1978).
28. S. P. Chashin, A. Z. Murzakhanova, I. P. Guzhova, *Sov. J. Opt. Tech.* 7, 9 (1986).
29. S. T. Kirsch, *Appl. Opt.* 20(12), 2085 (1991).
30. T. M. Ding, E. Garmire, *Appl. Opt.* 22(20), 3177 (1983).
31. F. Yang, J. R. Sambles, G. W. Braddberry, *J. Appl. Phys.* 78(4), 2187 (1995).

32. E. Pelletier, F. Flory, Y. Hu, *Appl. Opt.* 28(14), 2918 (1989).
33. R. Th. Kersten, *Vakuum-Technik* 23(1), 16 (1973).
34. H. P. Weber, F. A. Dunn, W. N. Leibolt, *Appl. Opt.* 12(4), 755 (1973).
35. P. A. Barnes, D. P. Schinke, *Appl. Phys. Lett.* 30(1), 26 (1977).
36. V. Oliver, J. C. Peuzin, *Appl. Phys. Lett.* 32(6), 386 (1978).
37. T. W. Hou, C. J. Mogab, *Appl. Opt.* 20(18), 3184 (1981).
38. F. Abeles, T. Lopez-Rios, *Opt. Commu.* 11(1), 89 (1974).
39. W. M. Robertson, E. Fullerton, *J. Opt. Soc. Am. B* 6(8), 1584 (1989).
40. P. A. Apanasevich, "Basic Conceptions of Cointeraction Light with Material". Nauka, Minsk, 1977.
41. E. Senguil, "Physics of Surface". Mir, Moscow, 1990.
42. A. Madan, M. P. Shaw, "The Physics and Applications of Amorphous Semiconductors". Academic Press, Boston, 1988.
43. V. L. Bonch-Bruevich, C. G. Kalashnikov, "The Physics of Semiconductors". Nauka, Moscow, 1977.
44. M. Cardona, Ed., "Light Scattering in Solids". Springer, Berlin, 1975.
45. G. Harbeke, Ed., "Polycrystal Semiconductor: Physical Properties and Application". Mir, Moscow, 1989.
46. A. V. Rakov, "Spectrophotometry of Thin-Film Semiconductors". Sov. Radio, Moscow, 1975.
47. A. Smith, "Applied Infrared Spectroscopy". Mir, Moscow, 1982.
48. D. P. Woodruff, T. A. Delchar, "Modern Techniques of Surface Science". Mir, Moscow, 1989.
49. N. Bloombergen, Ed., "Nonlinear Spectroscopy". Mir, Moscow, 1979.
50. "Spectroscopy of Glass Like System". Proc. A. Silin, Ed., Lativa State Univ., Riga, 1989.
51. V. A. Kiselev, V. V. Novikov, A. Cherdnichenko, "Exciton Spectroscopy of Semiconductor Surface". St.-Peterburg State University, St.-Peterburg, 1987.
52. A. A. Davidov. "IR-Spectroscopy of Oxide Surface". Nauka, Novosibirsk, 1984.
53. P. S. Kireev, "Physics of Semiconductors". V. School, Moscow, 1975.
54. Van Buren, "Defects in Crystals". In. Lit., Moscow, 1962.
55. A. Damask, J. Dins, "Point Defects in Metals". Mir, Moscow, 1966.
56. J. Kury, "Luminescence in Crystals". In. Lit., Moscow, 1962.
57. M. P. Shaskol'skaya, "Crystallography". V. School, Moscow, 1976.
58. S. Devidson, J. Levin, "Surface States". Mir, Moscow, 1973.
59. A. I. Golubov, "Quantum-Electron Theory of Amorphous Semiconductor". Ac. Sc. USSR, Moscow, 1963.
60. N. Mott, E. Devis, "Electron Process in Polycrystal Semiconductor". Mir, Moscow, 1974.
61. L. D. Landau, E. Mlifshic, "Theory of Field". Nauka, Moscow, 1973.
62. B. I. Stepanov, "Introduction in Modern Optics. The Basic Representations of an Optical Science on a Threshold XX Centuries". Nauka, Minsk, 1989.
63. B. I. Stepanov, "Introduction in Modern Optics. Quantum Theory of Interaction of Light with Substance". Nauka, Minsk, 1990.
64. M. Roberts, Ch. Makky, "Chemical of Gas–Metal Interface". Mir, Moscow, 1981.
65. V. M. Zolotarev, V. N. Morozov, E. V. Smirnova, "Optical Constant of Natural and Technical Media. Handbook". Chemistry, St.-Peterburg, 1984.
66. A. M. Goncharenko, V. P. Red'ko, "Introduction in Integrated Optics". Nauka i Techn., Minsk, 1975.
67. T. E. Plowman, S. S. Saavedra, W. M. Reichert, *Biomaterials* 19, 341 (1998).
68. T. Tamira, Ed., "Integrated Optics". Mir, Moscow, 1978.
69. J.-J. Kler, "Introduction in Integrated Optics". Sov. Radio, Moscow, 1980.

References

70. R. Hunsperger, "Integrated Optics: Theory and Technology". Mir, Moscow, 1985.
71. A. M. Goncharenko, V. A. Karpenko, "Theory of Optical Waveguides". Nauka i Techn., Minsk, 1983.
72. H. G. Unger, "Planar and Fiber Optical Waveguides". Mir, Moscow, 1980.
73. M. Born, E. Wolf, "Basic Concepts of Optics". Nauka, Moscow, 1973.
74. G. S. Landsberg, "Optics". Nauka, Moscow, 1976.
75. G. W. Pitt, F. R. Greller, *Thin Solid Films* 26(1), 127 (1975).
76. J. S. Wei, W. D. Westwood, *Appl. Phys. Lett.* V32 (12), 819 (1975).
77. V. E. Khomenko, A. A. Lipovsky, M. A. Alexandrov, *Instrum. Exp. Tech.* 3, 224 (1981).
78. P. K. Tien, R. Ulrich, *J. Opt. Soc. Am.* 60(10), 1325 (1970).
79. L. I. Derugin, N. Mmarchuk, V. E. Sotin, *Sov. Radio Electron.* 13(8), 974 (1970).
80. P. S. Chuny, *J. Appl. Phys.* 9(3), 887 (1976).
81. P. P. Herrman, *Appl. Opt.* 19(19), 3261 (1980).
82. A. N. Ageev, E. V. Mokrushina, A. C. Trifonov, *Tech. Phys.* 32(10), 2044 (1982).
83. E. M. Zolotov, V. A. Kiselev, V. A. Sychugov, *Sov. Phys.-Uspekhi* 112(2), 231 (1974).
84. L. M. Brekhovskikh, "Waves in Layer Media". Nauka, Moscow, 1973.
85. M. Adams, "Introduction in Theory of Optical Waveguides". Mir, Moscow, 1984.
86. G. V. Brandt, *Appl. Opt.* 14(4), 946 (1975).
87. P. S. Chany, *Opt. Acta* 23(8), 651 (1976).
88. M. Olivier, J. S. Danel, J. C. Penzin, *Thin Solid Films* 89(2), 295 (1982).
89. R. Torge, *Optic* 41(2), 212 (1974).
90. C. Este, W. D. Westwood, *J. Vac. Sci. Technol.* A5(4), 1892 (1987).
91. V. I. Borisov, A. I. Voitenkov, *Tech. Phys.* 51(8), 1668 (1981).
92. G. A. Muranova, V. S. Terpuchov, *News Acad. Sc. USSR* 45(2), 392 (1981).
93. N. Nourshargh, E. M. Starr, M. I. Fose, S. G. Jones, *Electron. Lett.* 21(18), 818 (1985).
94. A. Yariv, "Introduction in Optoelectronics". V. School, Moscow, 1983.
95. R. E. Kamke, "Ordinary Differential Equations. Handbook". Nauka, Moscow, 1989.
96. V. M. Agranovich, J. L. Mills, Eds., Surface Polaritons. Electromagnetic Waves on Surfaces and Interfaces". Nauka, Moscow, 1985.
97. A. Otto, *Z. Physik.* 216, 398 (1968).
98. E. Kretschmann, *Z. Physik.* 241, 313 (1971).
99. R. D. Olney, R. Romagnoli, *Appl. Opt.* 26(11), 2279 (1987).
100. E. Fontano, R. Pantell, M. Moslehi, *Appl. Opt.* 27(16), 3334 (1988).
101. B. S. Danilin, "Vacuum Technique in Integrated Schemes Manufacturing". Energiya, Moscow, 1972.
102. N. G. Einspruch, D. M. Brown, Eds., "Plasma Processing for VLSI". Mir, Moscow, 1987.
103. "Optoelectronic Materials". E. I. Givargizov, Ed., Mir, Moscow, 1976.
104. R. Th. Kersten, H. F. Mahlei, W. Ranscher, *Thin Solid Films* 28(2), 369 (1975).
105. "Science and Application of Solid Sputtering". E. S. Mashkova, Ed., Mir, Moscow, 1989.
106. Y. C. Chany, W. D. Westwood, Proc. Symp. Opt. and Acoust. Micro. Electron. New-York, 1974. Brooklyn, pp. 271–284, 1975.
107. B. A. Vishnyakova, K. A. Osipov, "Electron-Beam Technique for Thin Films and Chemical Connections Fabrication". Chemistry, Moscow, 1970.
108. V. K. Kiselev, V. P. Red'ko, *Sov. J. Quantum Electron.* 5(1), 134 (1978).
109. V. K. Kiselev, V. P. Red'ko, *Tech. Phys.* 49(4), 883 (1979).
110. G. T. Petrovsky, V. P. Red'ko, A. V. Khomchenko, *Tech. Phys.* 54(10), 2045 (1984).
111. V. N. Mogilevich, V. P. Red'ko, A. A. Romanenko, A. V. Khomchenko, *Tech. Phys.* 59(2), 91 (1990).
112. V. I. Anikin, L. N. Derugin, *Tech. Phys.* 47(10), 2163 (1977).
113. V. I. Anikin, L. N. Osadchev, *Microelectronics* 6(4), 369 (1977).

114. V. V. Vasil'ev, V. G. Pan'kin, V. P. Popov, *Autometry* 2, 22(1978).
115. R. W. Collins, M. Windischmann, J. M. Cavese, *J. Appl. Phys.* 58(2), 954 (1985).
116. J. E. Griene, A. H. Eltonkhy, *Surf. Interface Anal.* 3(1), 34 (1981).
117. B. R. Critcheby, P. R. Stevens, *J. Phys. D* 11(4), 481 (1978).
118. S. Toshikasn, M. Tadashi, *J. Cryst. Growth* 86(1–4), 423 (1988).
119. T. M. Galina, V. G. Volod'ko, E. S. Demidov et al., *Sov. Phys. Semicond.*. 27(8), 1379 (1993).
120. N. I. Aleshkevich, A. I. Voitenkov, V. P. Red'ko, *Sov. J. Quantum Electron.* 4(10), 2254 (1977).
121. V. A. Ganshin, Y. N. Korkishko, T. V. Morozova et al., *Tech. Phys.* 63(6), 166 (1993).
122. K. Mukesh, G. Vandana, G. DeBrabander et al., *IEEE Photonic. Tech. Lett.* 5(4), 435–438 (1993).
123. V. P. Red'ko, O. D. Shlyakhtichev, *Tech. Phys. Lett.* 23, 1414 (1978).
124. T. Kanata, Y. Kobayashi, K. Kubota, *J. Appl. Phys.* 62(7), 2989 (1987).
125. S. Nakahara, G. J. Fisanick, M. F. Yan, R. B. van Dover, T. Boone, *Appl. Phys. Lett.* 53(21), 2105 (1988).
126. A. A. Romanenko, A. B. Sdotsky, A. V. Khomchenko, Preprint No. 649, Institute of Physics of NASB, 1991.
127. S. Monneret, P. Huguet-Chant, F. Flory, *J. Opt. A* 2, 188 (2000).
128. H. Stark, Ed., "Applications of Optical Fourier Transforms". Radio i Svyaz', Moscow, 1988.
129. W. Demtroder, "Laser Spectroscopy. Basic Concepts and Instrumentation". Nauka, Moscow, 1985.
130. V. P. Klochkov, L. F. Kozlov, I. V. Putykevich et al., "Laser Anemometry, Remote Spectroscopy and Interferometry". Nauka, Kiev, 1985.
131. N. N. Evtikhiev, Y. A. Kupershmidt, V. F. Papulovsky et al., "Measurement of Electrical and Unelectrical Magnitudes". Energoatomizdat, Moscow, 1990.
132. Van der Zil, "Measurement Noise". Mir, Moscow, 1979.
133. A. F. Kotuk, Ed., "Measurements of Power Parameters and Characteristics of Laser Radiation". Radio i Svyaz', Moscow, 1981.
134. I. I. Anisimova, B. M. Glukhovsky, "Photoelectronic Multipliers". Sov. Radio, Moscow, 1974.
135. W. T. Tsang, Ed., "Lightwave Communications Technology. Photodetectors". Mir, Moscow, 1988.
136. A. Cozannet, J. Fleuret, H. Maitre et al., "Optics and telecommunications. Optical Transfer and Processing of the Information". Mir, Moscow, 1984.
137. A. V. Pavlov, A. I. Pavlov, A. I. Chernjakov, "Receivers of Radiation of Automatic Opto-Electronic Devices". Energy, Moscow, 1972.
138. V. V. Vasil'ev, I. Gurov, "Computer Processing of Signals in the Interferometry Applications". BHV, St.-Peterburg, 1998.
139. A. F. Koptuka, Ed., "Introduction in Measurement Technique of Physical Parameters of Lightwave Systems". Radio i Svjaz, Moscow, 1987.
140. M. D. Aksenenko, M. L. Baranochnikov, "Photodetectors". Moscow, Radio i Svjaz, 1987.
141. Y. R. Nosov, V. A. Shilin, "Physics of CCD". Nauka, Moscow, 1986.
142. J. Igan. "The Theory of Detection of Signals and Analysis of the Working Characteristics". Nauka, Moscow, 1983.
143. K.-H. Kunce, "Physical Measurement Methods". Mir, Moscow, 1989.
144. A. L. Mikaeljan, M. A. Ter-Mikaeljan, Y. G. Turkov, "Solid Lasers". Sov. Radio, Moscow, 1967.

145. B. I. Stepanov, Ed., "Modeling Methods of Optical Quantum Generators". Nauka i Techn., Minsk, 1968.
146. I. I. Pakhomov, O. V. Roszkov, V. I. Roszdestvich, "Optoelectronic Quantum Devices". Radio i Sv., Moscow, 1982.
147. M. Balkansky, P. Lammona, Eds., "Photonics". Mir, Moscow, 1978.
148. L. M. Kogan, Semiconductor lightdiodes. *Light technology* 6, 11 (2000).
148a. H. J. R. Dutton, "Understanding Optical Communications". IBM, http://www.redbooks.ibm.com, 1998.
149. Y. Rabek, "Techniques in Photochemistry and Photophysics". Vol. 2. Mir, Moscow, 1985.
150. J. I. Pankove, "Optical Processes in Semiconductors". Mir, Moscow, 1973.
151. M. I. Epstein, "Measurements of Optical Radiation in Electronics". Energoatomizdat, Moscow, 1990.
152. J. Redy, "Effects of Powerful Laser Radiation", pp. 14–15. Mir, Moscow, 1974 (in Russian).
153. J. Gudmen, "Statistical Optics". Mir, Moscow, 1988.
154. V. I. Smirnov, *Meas. Tech.* 3, 26 (1996).
155. A. Papulys, "The Theory of Systems and Transforms in Optics". Mir, Moscow, 1971.
156. Y. I. Ostrovsky, M. M. Butusov, G. V. Ostrovskaya, "Holographic Interferometry". Nauka, Moscow, 1977.
157. J. Keycecenta, Ed., "Optical Processing of Data". Mir, Moscow, 1980.
158. A. L. Mikaeljan, "Optical in Computer Science". Nauka, Moscow, 1990.
159. P. Huguest-Chantome, L. Escoubas, F. Flory, *Appl. Opt.* 41(16), 1 (2002).
160. V. P. Red'ko, A. A. Romanenko, A. B. Sotsky, A. V. Khomchenko, Patent Russian No. 2022247, G01N 21/43, 1994.
161. V. P. Red'ko, A. A. Romanenko, A. B. Sotsky, A. V. Khomchenko, *Sov. Tech. Phys. Lett.* 18(4), 14 (1992).
162. A. B. Sotsky, A. A. Romanenko, A. V. Khomchenko, Proc. Int. Conf. *CM NDT-95, Minsk*, p. 99 (1995).
163. I. U. Primak, A. A. Romanenko, A. B. Sotsky, A. V. Khomchenko, Proc. Int. Conf. Quantum Electron. Minsk, p. 102 (1996).
164. O. N. Kasandrova, V. V. Lebedev, "Processing of Recorded Data". Nauka, Moscow, 1970.
165. V. I. Mudrov, V. L. Kushko, "Processing Methods of Recorded Data". Sov. Radio, Moscow, 1976.
166. A. N. Lebedeva, E. A. Chernjavskogo, Eds., "Mathematic Optimization Methods in Computing Science". V. School, Moscow, 1986.
167. J. Hudson, "Statistics for Physics". Nauka, Moscow, 1970.
168. A. B. Sotsky, A. A. Romanenko, A. V. Khomchenko, I. U. Primak, *J. Commun. Technol. Electron.* 44(6), 640 (1999).
169. A. A. Romanenko, A. B. Sotsky, *Tech. Phys.* 68(4), 88 (1999).
170. V. Y. Vaydjalis, *Proc. VUZ Lithuania* 15(3), 123 (1979).
171. A. V. Khomchenko, in "Applied Optic Problems", p. 135 (V. P. Redko and V. A. Karpenko, Eds.). Mogilev State University, 2000.
172. A. A. Romanenko, A. B. Sotsky, A. V. Khomchenko, in Optics and Acoustics Inst. Physics, Minsk (1996) 71.
173. A. V. Khomchenko, A. B. Sotsky, A. A. Romanenko et al., *Tech. Phys. Lett.* 28(6), 467 (2002).
174. G. T. Petrovsky, V. P. Red'ko, A. L. Sorokovykh, A. V. Khomchenko, L. M. Steigart, *J. Appl. Spectroscopy* 42(1), 147 (1985).

175. A. V. Khomchenko, A. A. Romanenko, A. B. Sotsky et al., in "Applied Optic Problems", p. 156 (V. P. Redko and V. A. Karpenko, Eds.). Mogilev State University, 2000.
176. I. S. Malaschenko, V. P. Red'ko, L. E. Starovoytov, A. V. Khomchenko, *Surface* (in Russian) 1, 78 (1992).
177. J. Jounopulos, J. Lukovsky, Eds., "Physics of Hydrohenezirated Amorphous Silicon. B.2. Electronic and Oscillatory Properties". Mir, Moscow, 1988.
178. A. P. Sukhorukov, "Nonlinear Wave Interaction in Optics and Radiophysics". Nauka, Moscow, 1988.
179. "Absorption and Irradiation of Light by Quantum Systems". V. P. Grybkovsky, Ed., Nauka i Techn., Minsk, 1991.
180. Y. Herman, B. Wilgelmi, "Lasers of Supershort Light Pulses". Mir, Moscow, 1986.
181. "Laser Techniques and Devices for Measurements of Characteristics and Spectra of Substances". Proc. VNII PT&RTI, Moscow, 80 (1980).
182. "Techniques and Devices of Precision Spectroscopy". Proc. VNII PT&RTI, Moscow, 112 (1987).
183. B. Renby, Y. Rebek, "Photodestruction, Photooxidation, Photostabilization of Polymers". Mir, Moscow, 1978.
184. V. V. Lebedeva, "Techniques of Optical Spectroscopy". Nauka, Moscow, 1986.
185. I. M. Nagibina, V. K. Prokof'ev, "Spectral Devices and Spectroscopy Techniques". Nauka, Moscow, 1967.
186. A. C. Toporec, "Monochromators". Izd. Tehn. & Theor. Lit., Moscow, 1955.
187. V. V. Dubrovsky, A. I. Egorov, *Sov. Tech. Phys. Lett.* 19(19), 55 (1993).
188. S. L. Marpl, "The Digital Spectral Analysis and its Applications". Mir, Moscow, 1990.
189. B. R. Friden, Ed., "The Computer in Optical Research. Methods and Applications". Springer, Berlin, 1980.
190. P. I. Nikitin, *Sensor System* 12(1), 69 (1998).
191. V. A. Yakovlev, V. A. Sychugov, A. V. Tischenko, *Tech. Phys. Lett.* 8(11), 665 (1982).
192. M. Oliver, J.-C. Peuzin, J.-S. Danel, *Appl. Phys. Lett.* 38(2), 79 (1981).
193. E. Nitanai, S. Miyanaga, *Opt. Eng.* 35(3), 900 (1996).
194. A. V. Khomchenko, *Tech. Phys. Lett.* 27(4), 271 (2001).
195. V. V. Kisin, V. V. Sysoev, S. A. Voroshilov, *Tech. Phys. Lett.* 25(16), 54 (1999).
196. A. V. Khomchenko, D. N. Kostjuchenko, in "Applied Optic Problems", p. 149 (V. P. Redko and V. A. Karpenko, Eds.). Mogilev State University, 2000.
197. A. I. Voitenkov, V. A. Ereschenko, A. C. Borbitsky, *J. Appl. Spectrosc.* 63(2), 305 (1996).
198. V. I. Zemcky, Y. D. Kolesnikov, A. F. Novikov, *J. Opt.* 65(10), 16 (1998).
199. W. Lukosz, *Sensors Actuators* B29, 37 (1995).
200. A. Brandenburg, *Technisches Messen.* 62(4), 160 (1995).
201. P. Storm, "Theory of Probabilities. Mathematical Statistics". Mir, Moscow, 1970.
202. A. K. Mitropolsky, "Methods of Statistical Calculations". Nauka, Moscow, 1971.
203. A. G. Orlov, "Calculation Methods in Quantitative Spectral Analysis". Nedra, St.-Peterburg, 1977.
204. B. M. Schigalev, "Mathematical Processing of Recorded Data". Nauka, Moscow, 1969.
205. E. N. Dorokhova, G. V. Prokhorova, "Analytical Chemistry. Physico-Chemical Methods of the Analysis". V.school., Moscow, 1991.
206. Ph. M. Nellen, K. Tiefenthaler, W. Lukosz, *Sensors Actuators* 15, 285 (1988).
207. Ph. M. Nellen, W. Lukosz, *Sensors Actuators B* 1, 592 (1990).
208. Ph. M. Nellen, W. Lukosz, *Biosens. Bioelectron.* 6, 517 (1991).
209. W. Lukosz, D. Clerc, Ph. M. Nellen, *Sensors Actuators A* 25–27, 181 (1991).
210. Ph. M. Nellen, W. Lukosz, *Biosens. Bioelectron.* 8, 129 (1993).
211. Ph. M. Nellen, W. Lukosz, Ch. Stamm, P. Weiss, *Sensors Actuators B* 1, 585 (1990).

212. W. Lukosz, Ph. M. Nellen, Ch. Stamm, P. Weiss, D. Clerc, *Biosens. Bioelectron.* 6, 227 (1991).
213. W. Lukosz, Ch. Stamm, *Sensors Actuators A* 25, 185 (1991).
214. Ch. Stamm, W. Lukosz, *Sensors Actuators B* 11, 177 (1993).
215. W. Huber, R. Barner, Ch. Fattinger, J. Hubscher, H. Koller, F. Muller, *Sensors Actuators B* 6, 122 (1992).
216. R. Heideman, P. H. Kooyman, J. Greve, *Sensors Actuators B* 10, 209 (1993).
217. N. Fabricius, G. Gauglitz, J. Ingenhoff, *Sensors Actuators B* 7, 672 (1992).
218. L. M. Lechuga, A. T. M. Lenferink, R. P. H. Kooyman, J. Greve, *Sensors Actuators B* 24–25, 762 (1995).
219. E. F. Schipper, R. P. H. Kooyman, R. Heideman, J. Greve, *Sensors Actuators B* 24–25, 90 (1995).
220. T. Wink, S. J. Van Zuilen, A. Bult, W. P. Van Bennkom, *Anal. Chem.* 70, 827 (1998).
221. W. Lukosz, *Biosens. Bioelectron.* 6, 215 (1991).
222. D. C. Cullen, R. G. Brown, C. R. Lowe, *Biosensors* 3, 211 (1988).
223. P. B. Daniels, J. K. Deacon, M. J. Eddowes, D. G. Pedley, *Sensors Actuators* 15, 118 (1988).
224. M. T. Flanagan, R. H. Pantell, *Electron. Lett.* 20 (23), 968 (1984).
225. R. P. Kooyman, H. Kolkman, J. van Gent, J. Greve, *Anal. Chim. Acta* 213, 25 (1988).
226. B. Liedberg, C. Nylander, I. Lundstrom, *Sensors Actuators* 4, 299 (1983).
227. A. K. Nikitin, A. A. Tischenko, A. I. Cherny, *Zarub. Radioelectronik.* 10, 14 (1990).
228. R. E. Dessy, *Anal. Chem.* 61(19), 1079 (1989).
229. I. U. Primak, A. B. Sotsky, A. V. Khomchenko, *Tech. Phys. Lett.* 23(3), 46 (1997).
230. A. B. Sotsky, I. U. Primak, A. V. Khomchenko, A.V. Tomov, *Optical Quantum Electronics* 31(2), 191 (1999).
231. A. V. Khomchenko, E. V. Glasunov, I. U. Primak et al., *Tech. Phys. Lett.* 25(24), 11 (1999).
232. J. Homola, S. S. Yee, G. Gauglitz, Surface plasmon resonance sensors: Review. *Sensors Actuators B* 54, 3(1999).
233. V. P. Red'ko, A. V. Khomchenko, Patent USSR No1149202, G02F1/17, 1985.
234. A. V. Khomchenko, I. U. Primak, A. B. Sotsky, Proc. Int. Conf., Madras, India, 2, 1006 (1996).
235. A. B. Sotsky, I. U. Primak, *Doclady NASB* 42(2), 69 (1998).
236. V. F. Kiseljov, O. V. Krilov, "Electronic Phenomena in Adsorption and Catalysis". Springer, New York, 1986.
237. F. F. Volkenstein, "Electronic Processes on the Semiconductor Surface with Chemosorbtion". Nauka, Moscow, 1987.
238. F. Behsted, R. Enderlain, "Surfaces and Interfaces of Semiconductors". Mir, Moscow, 1990.
239. V. M. Aroutiounian, G. S. Aghababian, *Appl. Surf. Sci.* 135, 1 (1998).
240. S. V. Murav'eva, M. I. Bukovsky, E. K. Prokhorova, "Manual under the Testing of Impurity in Air of Working zone. Handbook". Chemistry, Moscow, 1991.
241. D. W. Pohl, *Thin Solid Films* 264(2), 250 (1995).
242. E. V. Glasunov, A. I. Voitenkov, I. U. Primak, et al., Proc. Int. Conf. CM NDT-98, Minsk. 291 (1998).
243. J. M. White, P. F. Heidrich, *Appl. Opt.* 15(1), 151 (1976).
244. A. V. Khomchenko, A. B. Sotsky, E. V .Glasunov, I. U. Primak, in "Applied Optic Problems". p. 166 (V. P. Redko and V. A. Karpenko, Eds.). Mogilev State University, Mogilev, 2000.
245. A. N. Grigorenko, A. A. Beloglazov, P. I. Nikitin, C. Kuhne, G. Steiner, R. Salzer, *Opt. Commun.* 174, 151 (2000).

246. W. Zhou, L. Cai, *Appl. Opt.* 37, 5957 (1998).
247. W. Zhou, L. Cai, *Meas. Sci. Technol.* 9, 1647 (1998).
248. S. Huang Peisen, *Appl. Opt.* 38(22), 4831 (1999).
249. J. Schemla, J. M. Ziss, Eds., "Nonlinear Optical Properties of Organic Molecules and Crystals". Vol.2. Mir, Moscow, 1989.
250. N. N. Rosanov, "Optical Bistability and Hysteresis in the Distributed Nonlinear Systems". Nauka, Moscow, 1997.
251. S. V. Gaponenko, "Optical Properties of Semiconductor Nanocrystal". University Press, Cambridge, 1998.
252. E. A. Andrushin, A. A. Bykov, *Sov. Phys.-Uspekhi* 154(1), 123 (1988).
253. H. Gibbs, "Optical Bistability: Controlling Light with Light". Academic Press, Orlando; Mir, Moscow, 1988.
254. L. A. Hornak and N. Y. Dekker, Eds., "Polymer for Lightwave and Integrated Optics: Technology and Applications". New York; Springer, Berlin, 1992.
255. U. Woggon, "Optical Properties of Semiconductor Quantum Dots". Springer, Berlin, 1996.
256. O. Levy, Y. Yagil, D. J. Bergman, *J. Appl. Phys.* 76(3), 1431 (1994).
257. Y. Q. Li, C. C. Sung, R. Inguva, C. M. Bowden, *JOSA B* 6(4), 814 (1989).
258. L. Banyai, Y. Z. Hu, M. Lindberg, *Phys. Stat. Sol. B* 3(8), 8142 (1988).
259. H. Haus, "Waves and Fields in Optoelectronics". Mir, Moscow, 1988.
260. I. R. Shen, "Principles of Nonlinear Optics". Nauka, Moscow, 1989.
261. V. M. Galitsky, V. F. Elesin, "Resonant Interaction of Electromagnetic Fields with Semiconductors". Energoizdat, Moscow, 1986.
262. N. Blombergen, "Nonlinear Optics". MIR, Moscow, 1966.
263. V. M. Fine, "Photons and Nonlinear Optics". Sov. Radio, Moscow, 1972.
264. A. V. Andreev, V. I. Emel'yanov, Y. A. Il'insky, "Cooperative Phenomena in Optics". Nauka, Moscow, 1988.
265. S. V. Gaponenko, *Sov. Phys-Semicond.* 30(2), 20 (1996).
266. D. N. Chigrin, A. V. Lavrinenko, D. A. Yarotsky, S. V. Gaponenko, *Appl. Phys. A* 68, 25 (1999).
267. V. P. Grybkovsky, "The Theory of Irradiation and Absorption of Light in Semiconductors". Nauka i Techn., Minsk, 1975.
268. P. I. Hadzhy, G. D. Shivarshina, A. H. Rotaru, "Optical Bistability in the Cogherence Excitons and Biexcitons System in Semiconductors". Stitsa, Kishinev, 1988.
269. P. A. Apanasevich, A. A. Afanas'ev, "Self-Diffraction and Compelled Dispersion of Light on Free Carriers in Semiconductors", Preprint No. 178, Institute of Physics of BAS, 1979.
270. D. A. B. Miller, *Laser Focus* 19(7), 61 (1983).
271. S. Schmitt-Rink, D. S. Chemla, D. A. B. Miller, *Adv. Phys.* 38(2), 89 (1989).
272. B. M. Bulakh, N. P. Kumin, V. P. Kunec et al., *Sov. Phys-Semicond.* 24(2), 254 (1990).
273. A. P. Bogatov, P. G. Eliseev, *Sov. J. Quantum Electron.* 12(3), 465 (1985).
274. V. P. Grybkovsky. Luminiscence, absorption and stimulation irradiation of light in semiconductors, in "Problems of Modern Optics and Spectroscopy". p. 203. Nauka i Techn., Minsk, 1986.
275. E. I. Rashba, M. D. Cterzha, Eds., "Excitons". Mir, Moscow, 1985.
276. R. K. Jain, *Opt. Eng.* 21(2), 199 (1982).
277. K. Klingshern, S. V. Gaponenko, *J. Appl. Spectrosc.* 56(4), 550 (1992).
278. S. A. Moskalenko, "Booze–Einstein Condensation of Excitons and Biexcitons". Stica, Kishinev, 1970.
279. T. Dekorsy, G. Segschneider, M. Tkim, S. Hunsche, H. Kurz, *Pure Appl. Opt.* 7, 313 (1998).

280. J. G. Winiarz, L. Zhang, M. Lal et al., *Chem. Phys.* 245, 417 (1999).
281. V. S. Gurin, M. V. Artem'ev, in "Chemical Problems of New Material and Technology Development". (V. V. Sviridov, Ed.). Belarus State Univ., Minsk, 1998.
282. V. V. Shashkin, *Avtometry* 3, 19 (1989).
283. V. T. Trofimov, Y. G. Selivanov, E. G. Chyzhevsky, *Phys-Semicond.* 30(4), 755 (1996).
284. A. N. Wais, L. V. Prokof'ev, *Phys.-Semicond.* 20, 160 (1986).
285. N. Y. Tabiryan, *Laser Focus World* 4, 165 (1998).
286. G. J. Bjorklund, G. T. Boyd, G. Carter et al., *J. Appl. Opt.* 26, 227 (1987).
287. M. Sheik-bahae, A. A. Said, E. W. Van Stryland, *Opt. Lett.* 14(17), 955 (1989).
288. M. Yin, H. P. Li, W. Ji, *Appl. Phys. B* 70(4), 587 (2000).
289. Y. J. Ding, C. L. Guo, G. A. Swartzlander, J. J. Khurgin, *Opt. Lett.* 15(24), 1431 (1990).
290. I. V. Tomov, B. Van Wonterghem, B. Dvornikov, et al., *J. Opt. Soc. Am.* 8(7), 1477 (1991).
291. T. Kobayashi, H. Uchiki, K. Minoshima, *Synthetic Metals* 28, D699 (1989).
292. S. P. Apanasevich, O. V. Goncharova, F. F. Karpushko, G. V. Sinicin, *J. Appl. Spectrosc.* 47, 200 (1987).
293. S. V. Gaponenko, V. P. Grybkovsky, A. G. Zimin, N. K. Nikienko, *Doklady ASB* 28(4), 318 (1984).
294. O. V. Goncharova, S. A. Tikhomirov, *Sov. J. Quantum Electron.* 22(4), 377 (1995).
295. I. Sarger, P. Segunds, F. Adamiets et al., *JOSA B* 11(9), 995 (1994).
296. W. S. Banyai, C. T. Seaton, G. I. Stegeman, M. O'Neill, et al., *Appl. Phys. Lett.* 54(6), 481 (1989).
297. A. S. Rubanova, Ed., "Manipulation of Wave Front of Laser Radiation in Nonlinear Media". Nauka i Techn., Minsk, 1990.
298. R. Horvas, J. Voros, R. Graf, et al., *Appl. Phys. B* 72, 441 (2001).
299. A. B. Sotsky, A. V. Khomchenko, L. I. Sotskaya, *Tech. Phys. Lett.* 20(8), 667 (1994).
300. H. Rigneault, F. Flory, S. Monneret, *Appl. Opt.* 34, 4358 (1995).
301. Y. B. Gideey, A. C. Trofimova, A. F. Prikhod'ko, et al., *Ukr. Phys. J.* 40(12), 37 (1995).
302. A. V. Tomov, *Sov. Tech. Phys. Lett.* 21(10), 25 (1995).
303. A. V. Tomov, A. V. Khomchenko, E. P. Kalutskaya, Proc. of XVII Int. Conf. on Coherence and Nonlinear Optics, Minsk, 2001.
304. A. V. Tomov, A. I. Voitenkov, A. V. Khomchenko, *Tech. Phys.* 68(2), 124 (1998).
305. Z. Sekkat, A. Knoesen, V. Y. Lee et al., *J. Phys. Chem.* 101, 4733 (1997).
306. M. Ivanov, T. Todorov, L. Nikolova et al., *Appl. Phys. Lett.* 66(17), 2174 (1995).
307. G. J. Lee, D. Kim, M. Lee, *Appl. Opt.* 34(2), 138 (1995).
308. L. N. Blinov, M. D. Bal'manov, N. S. Pochencova, *Sov. Tech. Phys. Lett.* 14(9), 86 (1988).
309. T. F. Masec, N. N. Smirnova, E. A. Smorgonskaya, V. K. Tikhomirov, *Sov. Tech. Phys. Lett.* 18(13), 46 (1992).
310. B. T. Kolomiec, K. P. Kornev, A. S. Kochemirovsky, L. V. Pivovarova, *Sov. Tech. Phys.* 52(12), 2424 (1982).
311. T. F. Masec, S. K. Pavlov, E. A. Smorgonskaya, E. I. Shifrin, *Sov. Tech. Phys. Lett.* 12(13), 802 (1986).
312. A. B. Sotsky, A. V. Khomchenko, L. I. Sotskaya, *Opt. Spectrosc.* 78(3), 502 (1995).
313. B. B. Boyko, N. S. Petrov, "Light Reflection from Amplify and Nonlinear Media". Nauka i Techn., Minsk, 1988.
314. A. Y. Winogradov, E. A. Smorgonskaya, E. I. Shifrin, *Sov. Tech. Phys. Lett.* 14(7), 642 (1988).
315. G. I. Stegeman, C. Y. Seaton, Nonlinear integrated optics. *Appl. Phys.* 58(12), R57 (1985).
316. V. M. Lubin, V. K. Tikhomirov, *Sov. J. Quantum Electron.* 19(4), 385 (1992).
317. V. M. Lubin, V. K. Tikhomirov, *Sov. Phys.-JETP Lett.* 51, 518 (1990).

318. V. M. Lubin, V. K. Tikhomirov, *Sov. Phys. Solids* 32, 1838 (1990).
319. A. I. Buzdugan, M. S. Iovu, A. A. Popesku, et al., *Sov. Tech. Phys. Lett.* 18(2), 6 (1992).
320. B. T. Kolomiec, T. F. Masec, S. K. Pavlov, *Sov. Phys. Semicond.* 12(8), 1590 (1978).
321. S. P. Apanasevich, O. V. Goncharova, F. F. Karpushko, G. V. Sinicin, *Bull. Acad. Sci. USSR, Phys. Ser.* 47(10), 963 (1983).
322. A. Andriesh, V. Chumash, *Pure Appl. Opt.* 7, 351 (1998).
323. R. Fisher, R. Muller, *Sov. J. Quantum Electron.* 16(8), 1723 (1989).
324. A. V. Khomchenko, E. V. Glazunov, *Tech. Phys.* 46, 1133 (2001).
325. L. Banyai, S. W. Koch, "Semiconductor Quantum Dots", Chap. 5. World Scientific, Singapore, 1993.
326. I. L. Krestnikov, N. N. Ledentsov, M. V. Maksimov et al., *Tech. Phys.Lett.* 23(1), 33 (1997).
327. A.V.Khomchenko, *Tech. Phys.* 42, 1038 (1997).
328. U. Z. Bubnov, M. S. Lur'e, F. G. Staros, G. A. Filaretov, "Vacuum Deposition of Films in Quasi-Closed Volume". Sov. Radio, Moscow, 1975.
329. I. P. Kalinkin, V. B. Aleskovsky, A. V. Simashkevich, "Epitaxial Films of II–VI Semiconductors". St.-Peterburg State Univ., St.-Peterburg, 1978.
330. G. T. Petrovsky, Ed., "Crystal Optical Material". Houme of Optics, Moscow, 1982.
331. T. Yodo, T. Koyoma, H. Ueda, K. Yamashita, *J. Appl. Phys.* 65(7), 2728 (1989).
332. Ya. S. Umansky, U. A. Skanov, A. N. Shanov, L. N. Rastorguev, "Crystallography, XRD and Electronic Microscopy". Metallurgy, Moscow, 1982.
333. L. I. Mirkin, "X-Ray Diffraction. Handbook". Phys.-Mat. St. Izd., Moscow, 1961.
334. Y. Kayanuma, *Phys. Rev. B* 38(14), 9797 (1988).
335. Al. L. Efros, A. A. Efros, *Sov. Phys.-Semicond.* 16(7), 1209 (1982).
336. K.V. Shalimova, "Physics of Semiconductors". Energy, Moscow, 1976.
337. A. V. Khomchenko, *Tech. Phys.* 45, 1505 (2000).
338. A. I. Ekimov, A. A. Onushchenko, *Sov. Tech. Phys. Lett.* 40(8), 337–340 (1984).
339. A. I. Ekimov, Al. L. Efros, A. A. Onushchenko, *Solid State Commu.* 56(11), 921 (1985).
340. S. Sh. Gevorkyan, N. V. Nikonorov, *Sov. Tech. Phys. Lett.* 16(13), 32 (1990).
341. V. A. Zul'kov, A. E. Kazachenko, L. N. Tvoronovich, *J. Appl. Spectrosc.* 50(1), 86 (1989).
342. M. Kull, J. L. Couraz, G. Mannelberg, "Research in Optics". Inst. of Opt. Res., Stockholm, 1989.
343. A. S. Borbitsky, A. I. Voitenkov, V. P. Red'ko, *Tech. Phys. Lett.* 22(13), 1 (1996).
344. A. V. Khomchenko, in "Applied Optic Problems" (V. P. Redko and V. A. Karpenko, Eds.). Mogilev State University, Mogilev, 2000.
345. S. M. Brehovskih, D. G. Galimov, U. P. Nikonov et al., *Phys. Chem. Glass* 6(3), 326 (1980).
346. N. M. Bobkova, A. K. Sinevich, *Phys. Chem. Glass.* 10(3), 337 (1984).
347. S. A. Kutolin, A. I. Neych, "Physical Chemistry of a Colour Glass". Chemistry, Moscow, 1988.
348. Ye. S. Yekimov, A. A. Onushchehko, V. A. Tcehosky, *Phys. Chem. Glass* 6(4), 511 (1980).
349. K. S. Binder, S. M. Oak, K. C. Rustagi, *Pure Appl.Opt.* 2(7), 345 (1998).
350. A. A. Borshch, V. I. Volkov, A. I. Mitckap, *Sov. J. Quantum Electron.* 22(4), 383 (1995).
351. T. Miyoshi, T. Miki, *Superlattice. Microstr.* 12, 243 (1992).
352. V. Ya. Grabovskih, Ya. Ya. Jenis, A. I. Yekimov, I. A. Kudryavtsev, et al., *Sov. Phys. Solid State* 31, 149 (1989).
353. D. I. Chepic, Al. L. Efros, A. I. Ekimov, M. G. Ivanov, V. A. Kharchenko, I. A. Kudriavtsev, T. V. Yazeva, *J. Lumin.* 47, 113 (1990).
354. J. Malhotra, D. J. Hagan, B. G. Potter, *J. Opt. Soc. Am. B* 8, 1531 (1991).
355. S. V. Gaponenko, I. N. Germanenko, V. P. Gribkovskii, M. I. Vasiliev, V. A. Tsekhomskii, in "Coexistence of Reversible and Irreversible Effects". *SPIE Proc.* 1807, 65 (1992).

356. K. S. Bindra, S. M. Oak, K. C. Rustagi, *Pure Appl. Opt.* 7, 345 (1998).
357. A. V. Khomchenko, *Appl. Opt.* 41(22), 4548 (2002).
358. A. V. Khomchenko, E. V. Glazunov, in "Nanostructure Materials. Fabrication and Properties". p. 143. Institute of Physics, Minsk, 2000.
359. A. V. Khomchenko, V. P. Red'ko, in Guides-wave optics. *Proc. SPIE.* 1932, 14 (1993).
360. A. V. Khomchenko, E. V. Glasunov, *Opt. Quantum Electron.* 34(4), 359 (2002).
361. A. V. Khomchenko, *Opt. Commun.* 201(4–6), 363 (2002).
362. D. N. Chigrin, A. V. Lavrinenko, D. A. Yarotsky, S. V. Gaponenko, *Appl. Phys. A* 68, 25 (1999).
363. A. V. Khomchenko, E. V. Glasunov, in "Physics, Chemistry and Applications of Nanostructures. Nanomeeting-99" (V. E. Borisenko, A. B. Filonov, S. V. Gaponenko and V.S. Gurin, Eds.) Minsk, 1999.
364. A. V. Khomchenko, in Reports of LON-99. Vilnius. 61 (1999).
365. L. S. Palatnik, P. G. Cheremskoy, M. Ya .Fuks, *"Porous in Films"*. Moscow, 1982.
366. K. Tanaka, Y. Ohtsuka, *J. Appl. Phys.* 49(12), 6132 (1978).
367. A. Andriesh, V. Chumash, *Pure Appl. Opt.* 7, 351 (1998).
368. R. V. Jansen, D. S. Wold-Kidane, O. F. Sankey, *J. Appl. Phys.* 64(5), 2415 (1988).
369. J. A. Wilson, V. A. Cotton, *J. Vac. Sci. Technol.* 3(1), 199 (1985).
370. B. I. Fuks, *JETP* 102(2), 555 (1992).
371. V. M. Abusev, E. I. Leonov, A. A. Lipovsky, S. E. Habarov, *Sov. Tech. Phys. Lett.* 13(20), 1268 (1987).
372. L. Roso-Franco. in "Optical bistability". Proc. of the Topical Meeting. Tiscon. AZ, USA. 2–4 December 1985, Springer, Berlin, XIV, 277 (1986).
373. A. V. Khomchenko, in Proc. of XVII Int. Conf. on Coherence and Nonlinear Optics Minsk. 4572, 305 (2002).
374. M. Allais, M. Gandais, *J. Appl. Crystallogr.* 23, 418 (1990).
375. L. C. Liu, S. Risbud, *Philos. Mag. Lett.* 61, 329 (1990).
376. L. E. Brus, *J. Chem. Phys.* 80, 4403 (1984).
377. P. D. Persans, L. B. Lurio, J. Pant, H. Yukselici, G. Lian, T. M. Hayes, *J. Appl. Phys.* 87(8), 3850 (2000).

Subject Index

absorption coefficient 16
   extinction 16
   attenuation 16
   nonlinear 159
absorption spectra 115, 118, 126

As_2S_3 films 173
adlayer 146
   evaluation of 146
adsorption 129

beam 54
   Gaussian 54, 64
   intensity 61, 64
   parameter 68
   splitter 63
   spatial distribution of 64
   width 54, 58, 64
buffer layer 105, 146, 165

chalcogenide semiconductor glasses 173
ChG 173
coherence 58
   length 59
"cutoff" condition 25, 26

defects 10
diffractogram 183
dispersion equations 25, 32
distribution 41
   spatial 41, 72
   intensity 41, 64
   mode field 53
   power level 63
   refractive index 90
   reflection coefficient 47, 87
Drude 19

Fourier transforms 57, 80

gap 72
guided mode 24, 77
   spatial Fourier spectroscopy of 77

impurity 8
   donor 9
   acceptor 9
incidence angles 24
intensity 17
Interface Effects 197
   distribution 53, 62, 65

Kerr media 157
Kramers–Kronig 19, 158

laser 60
leaky mode 37, 97
LED 51
lens 78, 81, 84
light source 50
Lorenz 20
losses 31
   measurement of 35

measurement error 83
microscopy 151
   waveguide 153, 156
mode 24
   guided 85
   leaky 37, 97, 99, 106
   imaginary part of 74
   propagation constant 23
   real part of 74
   excitation 83
model of free electrons 20
monochromator 117
multilayer structures 205

nonlinear optical constants 158, 167, 174–176
   measurement 164, 167

nonlinearity 158
    in azo-dye doped polymer films 169
    in multilayer structures 193
    in optical glass 186
    in semiconductor films 172
    in semiconductor-doped glasses 185
    origin of 159, 202
    parameter 158

optical loses 31, 93, 95

parameters 88
    of thin films 92, 104, 111, 174
permittivity of medium 14
photodetectors 43
    characteristics of 48, 49
    of the CCD-array type 48, 71
photomodulation spectroscopy 198
plasmon mode 37
polarization 13
polarizability 15
polarizer 84
Poynting vector 16, 17
power levels
    of impurity 8, 9
    of localized states 8, 12, 174
    of surface states 162, 185, 200
prism coupler 27, 64, 84
prism-coupling techniques 41, 83, 91
propagation of light 12
propagation constant 25, 41, 75
    imaginary part of 32, 74
    real part of 25, 74, 89

reflection coefficient 47, 99, 102
refractive index 15, 16
    anisotropy of 162
    complex 89
    nonlinear 158
rotary table 84

SCR 145
sensor 129
    gas 136, 149
    grating 134
    integrated-optics 131
    interferometer 133
    optimization of 150
    prism-based 136
    sensitivity of 134
    surface of 141
    surface plasmon wave 133
    time parameters of 149
source of radiation 47
    light-emitting diodes 51, 52
    laser 53, 56, 60
spatial distribution 41, 83
    intensity 72
states 5
    surface 11
    power 5
susceptibility 15
    nonlinear 158

total internal reflection 23

wave equation 37
wave number 87
waveguide 21
    fabrication of 39
waveguide spectroscopy 118, 129

XRD 183

ZnS 178
ZnSe 178–180, 182, 185